P9-CDG-392

Smithsonian

NATIONAL AIR AND SPACE MUSEUM

AN AUTOBIOGRAPHY

Part of the aviation model collection owned by the National Air and Space Museum is displayed in a glass case on the third floor of the National Mall building.

Smithsonian

NATIONAL AIR AND SPACE MUSEUM

AN AUTOBIOGRAPHY

EDITED BY MICHAEL J. NEUFELD AND ALEX M. SPENCER

FOREWORD BY JOHN H. GLENN, JR
AFTERWORD BY GENERAL JOHN R. DAILEY, DIRECTOR

NATIONAL GEOGRAPHIC

WASHINGTON, D.C.

C O N T E N T S

Museum founder Paul Garber stands at a National Air Museum model case in the Arts and Industries Building, probably in the late 1940s.

LEFT: John Glenn with his Mercury spacecraft, January 1962. RIGHT: John Glenn in the museum's Milestones of Flight gallery, 2001.

FOREWORD

The story of flight has been a central thread in my life. From my days flying for the Marines, to orbiting the Earth in *Friendship 7* in 1962, to returning to space aboard the space shuttle *Discovery* in 1998, I have experienced and witnessed the progress of aerospace technology for half its history. Equally satisfying to me has been my association with the National Air and Space Museum. This great institution houses and displays many of the milestone vehicles of flight, beginning with the Wright brothers' 1903 Flyer.

BY JOHN H. GLENN, JR.

But the museum does much more. It also preserves and shares the heritage of flight through the personal stories of the men and women who created and advanced this world-changing technology. The National Air and Space Museum is not just a collection of amazing machines. It is a showcase for stories of great individual and collective human achievements. I am honored to be part of those stories.

The Smithsonian's connection to flight began with the birth of the Institution, first headed by Joseph Henry, a physicist, balloon enthusiast, and sky-watcher. Samuel Langley, the third Secretary of the Smithsonian and a contemporary of the Wright brothers, headed a team that designed, built, and flew the first genuinely successful steam-powered model airplane but failed to fly a full-scale piloted version. His successor, Secretary Charles Walcott, played a key role in the creation and early history of the National Advisory Committee for Aeronautics, the predecessor of today's NASA. Perhaps just as important, it was Walcott who hired a 20-year-old aviation enthusiast, Paul Edward Garber, in 1920.

Today's National Air and Space Museum began as a gleam in young Paul Garber's eye. At the time of his death in 1992, he had been coming to work at the Smithsonian for 72 years. The long list of historic aircraft and spacecraft that he shepherded into the national collection ranges from Charles Lindbergh's 1927 *Spirit of St. Louis* to the Bell X-1 *Glamorous Glennis* that carried Chuck Yeager through the sound barrier in 1947 to the Mercury capsules in which I and the other American astronauts first ventured into space. You can thank Paul Garber for the survival of a great many of the aerospace treasures that delight, educate, and inspire the millions of visitors who pour through the doors of the museum on the mall and the Steven F. Udvar-Hazy Center at Dulles Airport each year.

When the museum opened in 1976, it was far more popular and successful than anyone had anticipated. Record crowds beat a path to the museum from the first day the doors opened, and they have kept coming unabated year after year, drawn by a fascination with the story of flight and the opportunity to see so many historic aircraft and spacecraft in one place. At a deeper level, I think the museum makes visitors feel proud to be human—to be members of a species that has traveled from a lonely beach at Kitty Hawk, North Carolina, to the dusty surface of the moon in less than seven decades, and is now sending robot explorers to the planets, their moons, and beyond the limits of the solar system.

The story of the growth of the national aerospace collection, the fight for facilities to preserve and display the machines that wrote history in the sky, and the generations of men and women who made the museum what it is today makes for interesting reading. While it is not a story of unalloyed success, the stumbles that have occurred along the way only serve to underscore the fact that the museum is an important place that matters to Americans and to many others around the world. With that said, I invite you to sit back and enjoy. Leafing through this book is the next best thing to visiting the museum.

INTRODUCTION

The Smithsonian Institution's National Air and Space Museum has been, ever since the building on the National Mall opened in 1976, among the top five most visited museums in the world, and it has often been ranked number one. The staff here felt that a history of this great institution was long overdue, so we decided to do it ourselves, as an "autobiography."

BY MICHAEL J. NEUFELD & ALEX M. SPENCER

Like any autobiography, or internally written history, what we present here is clearly our view of ourselves, not one that an uninvolved historian might write. Nonetheless, we feel that the material we present here is not uncritical and is based on a solid foundation of archival research and personal experience with the museum.

The book is organized into five major chapters, beginning with the Smithsonian's decades of engagement with aeronautics before World War I; proceeding to curator Paul Garber's amassing a collection of historic airplanes between the wars; on to the long struggle to construct a museum on the National Mall; and then to two parallel stories, one on the mall museum since its opening, and the other on the origins, construction, and life of our second major site, the Steven F. Udvar-Hazy Center, which opened in 2003 and was significantly expanded in 2010.

Between these main chapters are four interchapter features, longer discussions of decade-spanning stories of the Smithsonian's engagement with flight: the controversy between the Wright brothers and the Institution, which delayed the arrival of the historic 1903 Flyer until 1948; the Smithsonian's involvement with rocket pioneer Robert Goddard between the world wars; the long history of the museum's engagement with NASA; and the fascinating history of how the B-29 *Enola Gay*, the aircraft that dropped the first atomic bomb on Hiroshima, is intertwined with the history of the museum. Interspersed in the chapters are shorter sidebars that shed light on other key artifacts and the people who worked here, along with a rich array of photographs, many of them never before published.

The editors above all would like to thank Melissa Keiser for her indefatigable efforts in editing and supplying the images for this book and Tom Crouch, who not only drafted more prose than anyone else but also effectively acted as the third editor. We are also grateful to Bob van der Linden, Dom Pisano, Ted Maxwell, and Dik Daso for their major contributions to the text, and Dorothy Cochrane, Jeremy Kinney, Christopher Moore, and David DeVorkin for short pieces. We also wish to thank Marilyn Graskowiak for her research into historical photographs, and Mark Avino, Eric Long, and Dane Penland for their original photography. Patricia Graboske, museum publications officer, played a crucial role in the agreement with National Geographic Books and has been a valuable liaison with the staff there. Finally, we wish to thank Peter Jakab, associate director for collections and curatorial affairs, and John R. Dailey, director, for their support of, and contributions to, this project. We would also like to thank a true friend of the museum, John Glenn, for his willingness to contribute the Foreword.

From outside the museum, we would also like to acknowledge Jim Mahoney, former chief of exhibits at the National Air and Space Museum in the 1960s and 1970s. He provided photographs and advice, as well as his recollections of things in the Arts and Industries and Aircraft buildings. For their invaluable assistance in archival research, the authors would like to recognize Pamela Henson and Ellen Alers of the Smithsonian Institution Archives and Colin Fries and John Hargenrader of the National Aeronautics and Space Administration's History Division. Alan Meyer of Auburn University did original research on the museum's fellowship program that was helpful to the authors of Chapter 4.

At National Geographic Books, we very much enjoyed working with Barbara Brownell Grogan, Susan Tyler Hitchcock, Marianne Koszorus, Sam Serebin, Bronwen Latimer, and Marshall Kiker. We thank them for their editorial insight, visual design sense, and eye for a striking photograph. It is a much better book because of their participation.

The 1903 Wright Flyer hangs over the Apollo 11 command module *Columbia* in the museum's Milestones of Flight gallery in 2002.

The Smithsonian's Aircraft Building, nicknamed the Tin Shed, shown here in 1935, stood behind the Castle on Washington's National Mall.

The museum displayed the U.S. Navy Curtiss NC-4 flying boat on the Washington Mall in summer 1969 to celebrate the 50th anniversary of the first transatlantic flight.

The National Mall building of the National Air and Space Museum

The National Aviation and Space Exploration Wall of Honor stretches outside the Steven F. Udvar-Hazy Center, shown here in 2004.

1 FLIGHT AND THE SMITHSONIAN

TOM D. CROUCH

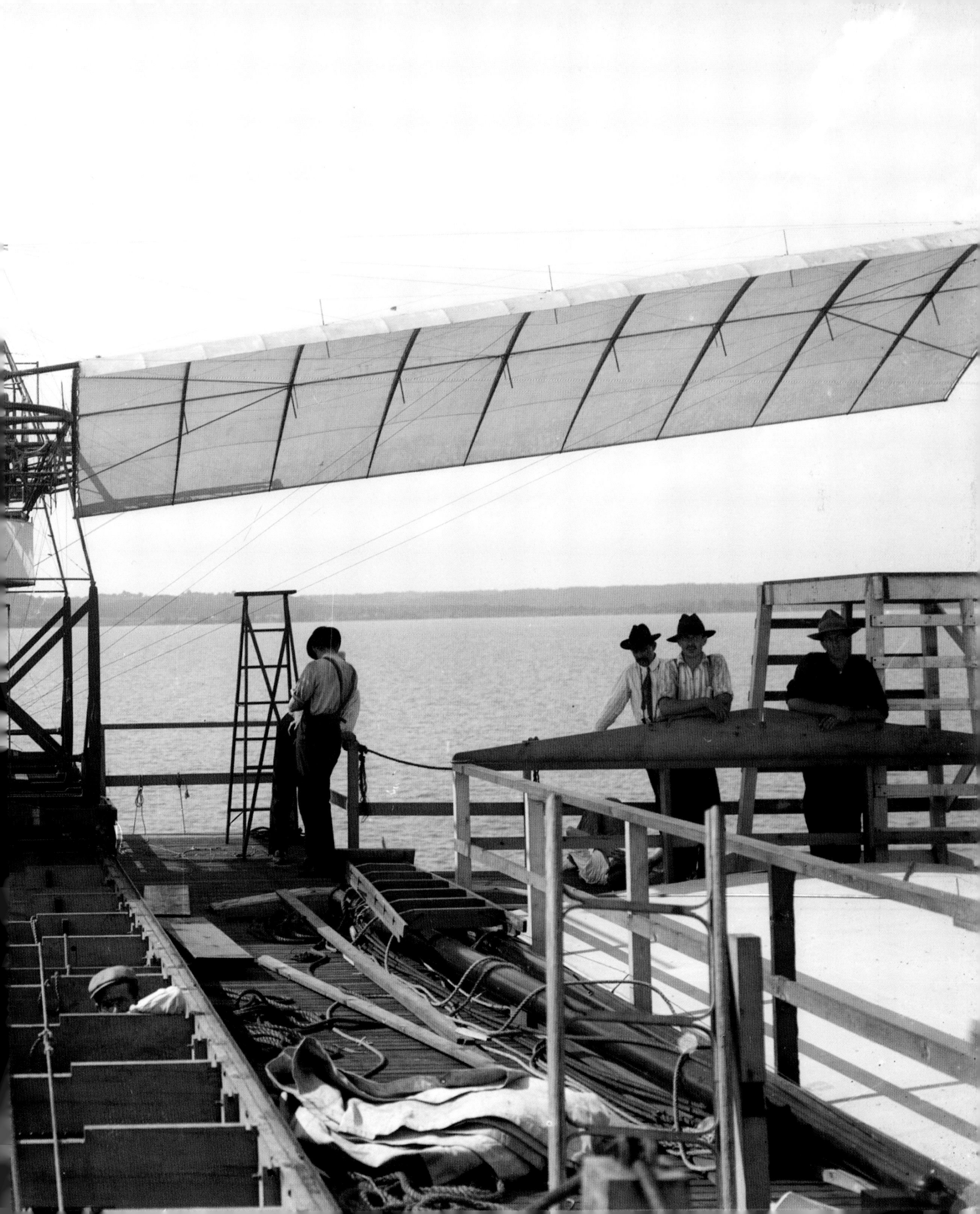

The millions of visitors *who pass through the*

in Washington, D.C., each year come to see the real thing, the actual aircraft and spacecraft that shaped history—from the world's first airplane to the backup hardware for the latest robot spacecraft sent forth to explore another world. Few if any of those visitors realize that aerospace history was made here, on the very site of today's museum on the National Mall, and 42 years before the first flight of an airplane. / It was June 16, 1861, and the Civil War had been under way for just two months. The first major battle of the war, which would take place near a quiet stream called Bull Run, 30 miles southwest of Washington, was still a little over a month away. At the *doors of the National Air and Space Museum*

time, the Columbia Armory stood where the museum is now located, east of Washington's Seventh Street, at the extreme southeastern tip of the 52-acre plot of land then known as the Smithsonian Grounds. The generating plant for the Washington Gas Light Company was immediately east of the armory, along with a large domed storage tank for the coal gas produced by the plant. It was that useful combination—the available work space at the armory and the gas company next door—that led Joseph Henry, then Secretary of the Smithsonian Institution, to instruct Thaddeus Sobieski Constantine Lowe to inflate his balloon on the site.

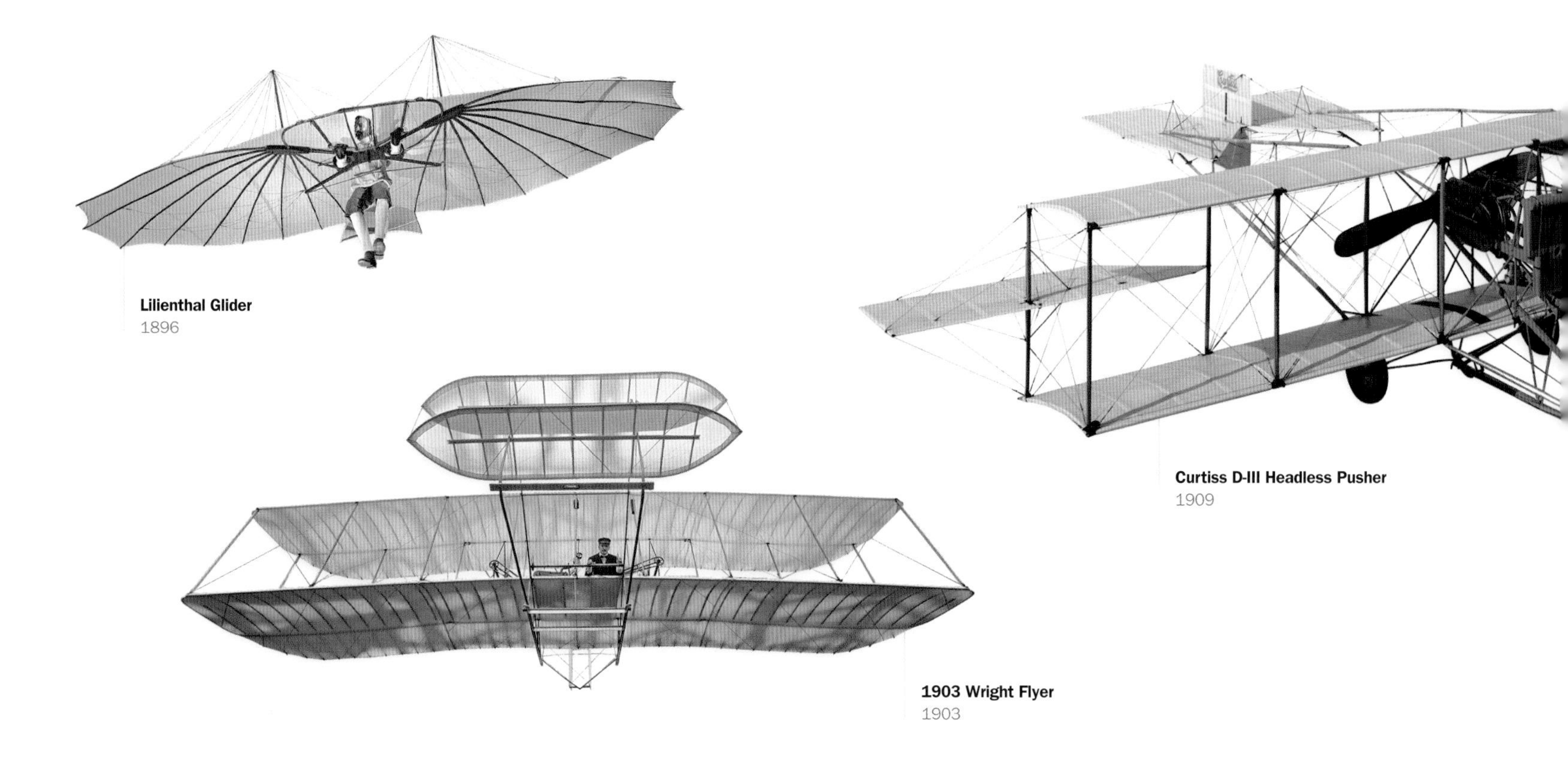

Lilienthal Glider
1896

Curtiss D-III Headless Pusher
1909

1903 Wright Flyer
1903

FROM THE MUSEUM'S COLLECTION

1890s TO 1920s

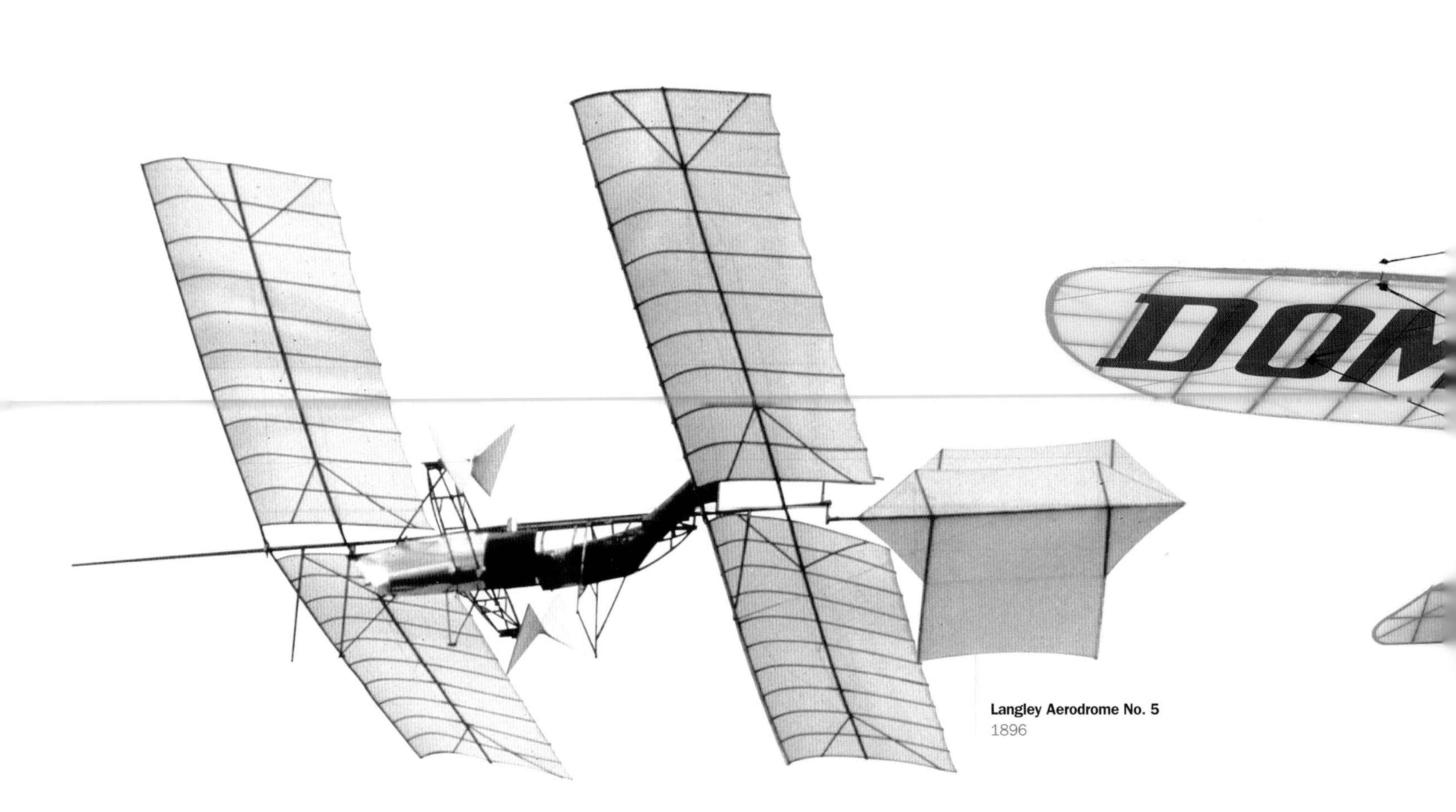

Langley Aerodrome No. 5
1896

Wright EX *Vin Fiz*
1909

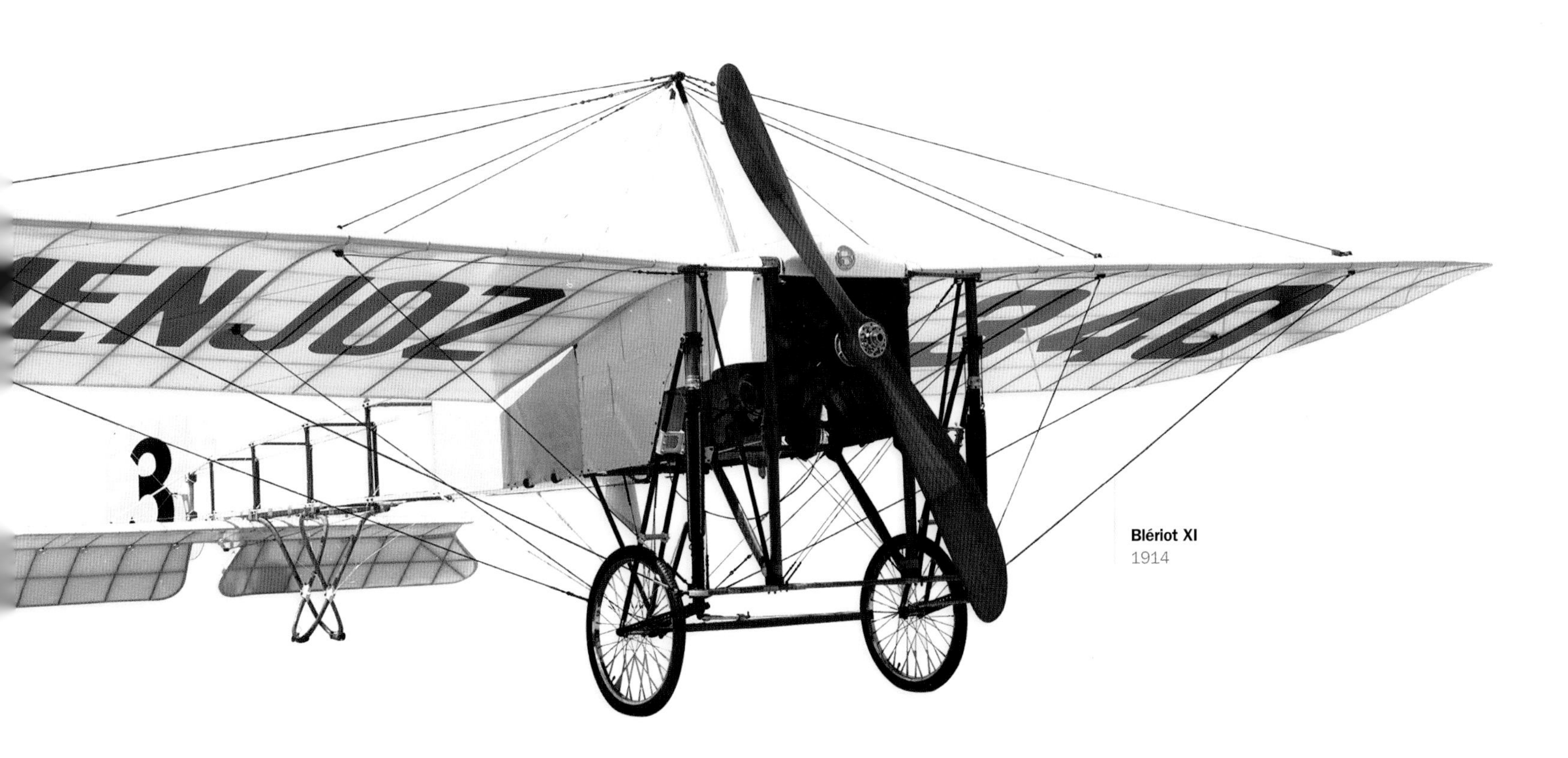

Blériot XI
1914

Spad XIII *Smith IV*
1917

Sopwith 7F.1 Snipe
1918

Voisin Type 8
1916

de Havilland DH-4
1918

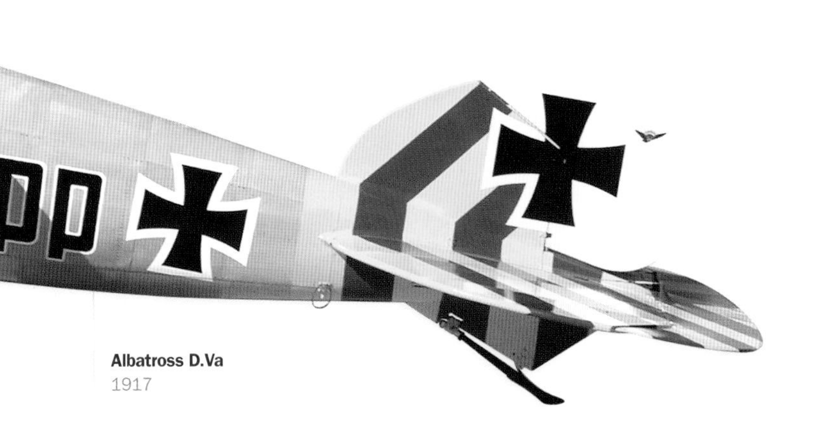

Albatross D.Va
1917

Verville-Sperry M-1 Messenger
1920

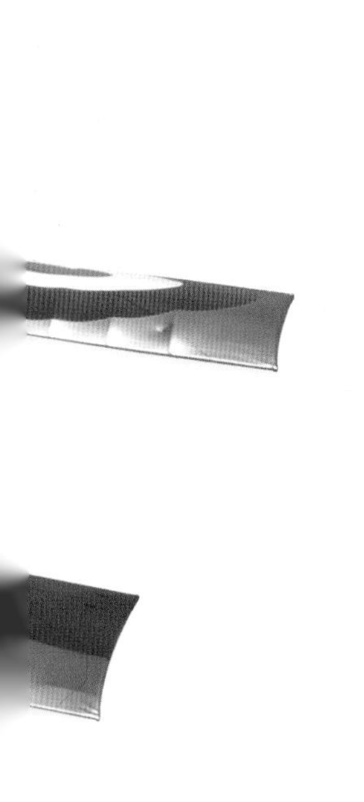

Halberstadt CL.IV
1917

Lowe made his first flight in 1857 and burst into the headlines two years later with a giant balloon, exhibited in New York and Philadelphia, with which he hoped to fly the Atlantic. When that plan fell through, and on the advice of the Smithsonian's Joseph Henry, his scientific consultant, Lowe flew from Cincinnati to Unionville, South Carolina, aboard the balloon *Enterprise,* on April 19–20, 1861. Landing a week after the firing on Fort Sumter, the New Hampshire aeronaut was taken into custody by newly minted Confederates and released only after locals recognized his face from accounts of his transatlantic plans published in the illustrated national newspapers of the day.

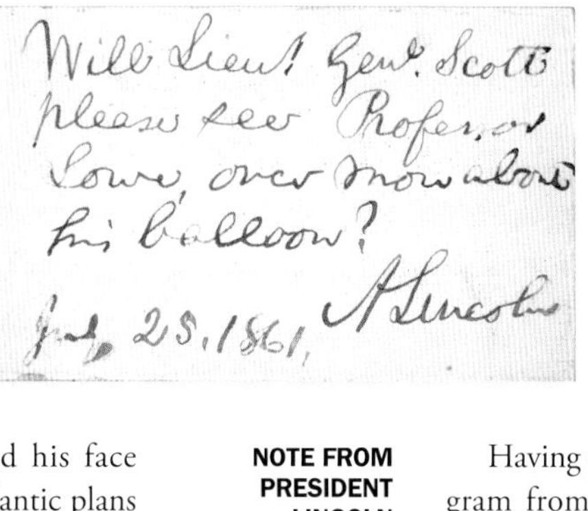
Will Lieut. Genl. Scott
please see Professor
Lowe, once more about
his balloon?
July 25, 1861, A. Lincoln

NOTE FROM PRESIDENT LINCOLN 1861
First presidential mention of aircraft in service to U.S. government

OPPOSITE: Thaddeus S. C. Lowe (1832–1913), founder and head of the Union Army balloon corps

PAGES 18–19: Smithsonian workmen look on as the partially assembled Langley Aerodrome sits in the catapult. The rear wings have yet to be attached. The photo was probably taken not long before the first launch attempt on October 7, 1903.

Urged on by Joseph Henry, and armed with letters of introduction to political figures in the new Lincoln Administration, Lowe packed up the *Enterprise* and traveled to Washington. Henry took the aeronaut to meet with President Abraham Lincoln on June 11, assuring Lincoln that Lowe was a leader in the field. Henry then arranged for the gas company to inflate the balloon and ordered William Jones Rhees, the Smithsonian's chief clerk, to provide Lowe with a crew of men to assist with the operation. The War Department would eventually reimburse the Smithsonian $250 for expenses related to the demonstration flight.

With Smithsonian support, Lowe made a series of tethered ascents from the area in front of the armory. The most important of those flights came on June 16, 1861, when he rose 500 feet, accompanied by telegrapher Herbert Robinson and George Burns, supervisor of the telegraph company. With a clear view of the nation's capital, Lowe sent a telegram to the White House, stating: "This point of observation commands an area near fifty miles in diameter. The city with its girdle of encampments presents a superb scene. I have pleasure in sending you this first dispatch ever telegraphed from an aerial station and in acknowledging indebtedness to your encouragement for the opportunity of demonstrating the availability of the science of aeronautics in the military service of the country."

Having dispatched the first ever telegram from the air, Lowe ordered his balloon winched down to the ground and walked to the White House, where he again met with the President. The two stayed up late discussing the military potential of balloon reconnaissance. Lincoln insisted that Lowe spend the night at the White House so that the pair could continue their discussion over breakfast. That first flight from the spot where the National Air and Space Museum would one day rise was the first demonstration of aerial reconnaissance in American history, and it led directly to the establishment of a balloon corps for the Union armies, the nation's first military aviation unit.

ROOTS

When T. S. C. Lowe rose above the Columbia Armory in the early summer of 1861, the Smithsonian was just 15 years old, having been created by an act of Congress on August 10, 1846, to meet the stipulations of the will of James Smithson, the illegitimate son of the Duke of Northumberland. A talented researcher who contributed to the fields of chemistry, mineralogy, and geology, he was the youngest member of the Royal Society at the time of his election in 1787.

The childless Smithson was also a shrewd investor, and he bequeathed his considerable fortune to his nephew. In the event that the young man should die without heirs, either legitimate or illegitimate, the money was to go "to the United States of America, to found at Washington, an establishment for the increase and diffusion of knowledge among men." Unfortunately for the nephew, but fortunately for the rest of us, that is exactly what happened.

After considerable debate, Congress agreed to accept the gift ($508,318 worth of gold sovereigns) in 1836. Precisely how to achieve Smithson's goal of promoting "the increase and diffusion of knowledge among men" occasioned yet another extended debate. Would the Smithsonian Institution function as a university, a library, a center for agricultural research, a museum? In the end, the decision was to build an imposing building and allow the first chief executive of the new organization, Joseph Henry, to shape the program. Under his leadership, the Smithsonian slowly gained distinction as both a research establishment and as America's center for the collection and distribution of scientific information.

Joseph Henry's interest in flight predated the establishment of the Smithsonian. A native of Albany, New York, he worked as a watchmaker, silversmith, and civil engineer, earning the

money to attend the Albany Academy, where he would remain as professor of mathematics and natural philosophy following his own graduation in 1826. His pioneering work on magnetism and electricity helped lay the foundation for the invention of the telegraph, telephone, and electric motor. By 1846, when he accepted the appointment to head the newly created Smithsonian Institution, Henry was generally regarded as the most important American scientist since Benjamin Franklin.

Joseph Henry developed an early appreciation for the potential of aeronautics. He witnessed his first balloon ascent on May 2, 1835, when he and a friend joined thousands of spectators crowding the neighborhood around Ninth and Green Streets in Philadelphia. This was the very first ascent of John Wise. Wise had made cabinets and pianos in his native Lancaster, Pennsylvania, before launching his career as a balloonist. Over the next two decades, he would emerge as the best known and most successful of American aeronauts. Between 1835 and the time of his disappearance during a flight over Lake Michigan in the fall of 1879, he would complete 1,450 ascents. He was an innovator as well, introducing the ripping panel, a lightly stitched section of fabric that could be pulled loose to allow the gas to escape quickly upon landing.

With Henry looking on, Wise got his aeronautical career off to a shaky start. Having already canceled his ascent once because of bad weather, he faced the prospect of sparking a riot if he did not take to the air as promised on May 2. When the balloon was released, it rose to an altitude of 40 feet, then dropped back to earth. Wise dumped ballast, climbed over a four-story house, and returned to earth a second time. With the hostile crowd moving ominously toward the landing spot, Henry later reported to his wife that Wise "then threw out his instruments, his coat, his hat, and every article." At that point, he finally rose to an altitude of one mile and flew off "until he was lost by the hills of New Jersey."

Henry's next encounter with a balloon came on July 24, 1837, when he witnessed one of the most publicized aeronautical tragedies of the era while visiting London. The famed English aeronaut Charles Green, accompanied by his friend Edward Spencer, agreed to carry the painter Robert Cocking aloft from the Vauxhall pleasure gardens to test a new parachute design. Looking something like an inverted umbrella, the device featured three metal hoops to maintain the shape of the fabric and weighed some 223 pounds. "The parachute descended . . . rapidly towards the Earth," Henry explained to his wife, "broke in its desent [*sic*] and precipitated the unfortunate adventurer head long to the ground."

The tragedy did nothing to discourage Henry's interest in the possible scientific applications of the balloon. Convinced that a better understanding of meteorology and improved weather forecasting would be of enormous value to Americans, he created a far-flung network of correspondents who mailed and telegraphed up-to-the-minute weather data to the Smithsonian. Henry created a large map on which to track weather changes and began to develop a broad understanding of America's climate, which he shared with other scientists.

OLDEST KNOWN PHOTO OF AN AIRCRAFT June 1857 Aeronaut John Steiner inflates his balloon at Erie, Pennsylvania.

OPPOSITE: Joseph Henry (1797–1878), first Secretary of the Smithsonian Institution, promoted the use of balloons for both scientific research and military purposes.

The research brought him into contact with balloonists, men who were experienced at riding the winds. John Wise, whom Henry had seen fly from Philadelphia a decade before, wrote to the Secretary in the spring of 1849, inquiring as to the possibility of using experiments carried aboard a balloon to prove or disprove the hollow earth theory originally proposed by Ohio land speculator John Cleves Symmes. As gently as possible, the Secretary explained to Wise that he could not imagine any such experiment.

Far from dismissing Wise for his interest in outmoded pseudoscientific theories, Henry recognized the aeronaut as a practical student of meteorology who had confirmed his own suspicion that an upper level current of air moved across the United States from west to east. At Henry's invitation, Wise visited the Smithsonian in April 1858. The nation's most experienced aerial traveler, Wise told Henry "all I knew about thunderstorms and atmosphere phenomena, so far as I had observed and experienced the workings of nature, both in and outside of the clouds." Having sparked the Secretary's curiosity, the aeronaut proposed to build a balloon expressly designed for atmospheric research, "to be conducted under the auspices of this learned philosopher."

Wise later recalled that the Secretary was especially interested in "the practical idea of appropriating natural electricity as a motor for engines." A natural force, "capable of pulverizing rocks, splitting up trees, knocking down masonry, and ploughing up the | *to page 32* |

dollars. Should I meet with much delay in getting this amount it will probably defeat the object for which I have been labouring for many years, and will consequently put me to much distress. Again, asking your pardon for troubling you. Knowing as I do. that in addition to your labors at the Smithsonian Institution. that much of your time is occupied in rendering valuable scientific service to the General Government.

I remain with great respect.

Your ever Obd't Servt

T. S. C. Lowe

Aeronaut.

No 1617 Race St.

By the summer of 1863, Thaddeus Lowe was losing patience with the Union Army officials who refused to reimburse him for his expenses. In this letter of July 15 he seeks Smithsonian Secretary Joseph Henry's help in resolving the matter.

20624 July 24/63 487

Philada July 15th 1863

To.

Professor Joseph Henry
Secretary of the Smithsonian Institut[e]
Washington D.C.

Dear Sir

I beg pardon for troubling you with my affairs connected with the Government but inasmuch as the first operations of Balloons for Milita[ry] purposes were under your immediate supervision, and you being acquainted with the fact that the experiments were made with my own machinery and subsequently used in the field by order of Captain A. W. Whipple late Genl Whipple now dead and from whom I can get no assistance. I hope that you will find it consistent to furnish to the Hon. Secretary of War such a statement as will satisfy him of the truthfulness of my claims. In order that you may know what my claims are and judge of the correctness I enclose them with this letter.

| from page 29 | earth," he noted, "wants only to be properly harnessed to make it subservient to human purposes." In fact, Henry doubted the practicality of harnessing lightning. At the same time, he assured Wise, "I think the investigations you propose are more interesting than any in the whole domain of meteorology and I am acquainted with no person better qualified than yourself to undertake them."

Henry immediately agreed to furnish the lifting gas and associated equipment required to inflate a balloon, as well as the scientific instruments to be carried aloft. Wise completed work on the new balloon in May 1859. It was named *Smithsonian* and bore the motto *"Pro Scientia et Arts."* He first flew the balloon from the Centre Square of Lancaster that month, rising into the teeth of a thunderstorm; he wrote, "Having noticed some remarkable phenomena during this voyage, such as an incipient thundercloud—the formation of a water-spout hanging down from this cloud—the increase of the cloud into a regular thundergust, and while sailing in the trail of the storm, that is in the rear of the ascending vortex, encountering large drops of rain dashing against the balloon and scintillating fire as they struck the balloon, it is needless to say I hurried down upon that demonstration."

Upon reading Wise's report of the inaugural voyage of the balloon *Smithsonian,* Henry informed the balloonist that he would have a few weeks of vacation in the summer of 1859 and suggested that he "would be pleased to make some of the experiments with you which we contemplated last summer."

CARTE DE VISITE Circa 1870 A calling card from the noted American balloonist John Wise (1808–1879), who sought to interest the Smithsonian in ballooning

OPPOSITE: Thaddeus Lowe goes aloft aboard the balloon *Intrepid* to observe Confederate activity during the Battle of Fair Oaks, May 31–June 1, 1862.

It was not to be, however. Wise spent the summer of 1859 preparing to fly a balloon from St. Louis to the Atlantic coast, while Henry, whatever his dreams of aerial adventure, continued to struggle with his administrative burdens. Addressing a scientific meeting in June 1859, however, the Secretary expressed full confidence that Wise's dream of riding that great eastbound current of air across the Atlantic "was by no means improbable." He considered the balloon to be "a very important instrument in meteorology and noted that the "observations of Mr. Wise have been of very great value."

Henry continued to extol the potential of John Wise and his balloon to advance meteorological knowledge to colleagues for some years to come. The scientist and the aeronaut would remain friends for almost three decades. In his last letter to Wise in July 1874, the Secretary commented on the balloonist's continued plans for a transatlantic flight, this one sponsored by the *Daily Graphic* newspaper. While assuring Wise that he remained convinced that such a trip was possible, he was also quick to point out the risks. "While I would be delighted to learn that you had successfully accomplished the feat," he explained, "I would prefer that someone in whom I am less interested were subjected to the risk."

Wise clearly appreciated Henry's support over the years. He dedicated his 1873 magnum opus, *Through the Air,* to the Secretary, "as a tribute of respect and admiration."

Henry's other aeronautical friend, T. S. C. Lowe, also dreamed of ballooning across the Atlantic. In December 1860, the Secretary received a letter from a group of Lowe's wealthy Philadelphia supporters asking his opinion on the feasibility of such a voyage. Henry assured them that strong currents blowing from west to east across the ocean should enable a well-designed and -constructed balloon to make the trip. He did, however, advise a trial shakedown flight from the middle of the continent to the coast. As a result, Lowe made his 1861 flight from Cincinnati to South Carolina. When the aeronaut subsequently asked if the Smithsonian would be willing to invest in his transoceanic venture, Henry explained that there was no money for any balloon project and that "the political events of the country have rendered it necessary for us to be exceedingly cautious in attempting any new enterprises."

He suggested that there was money to be made touring the major cities of the Union offering passenger rides in a tethered balloon. Almost as an afterthought, he added that Lowe and his balloon "might even be of advantage to the Government in assisting their reconnaissance of the district of country around Washington." That was all the invitation Lowe needed. Gathering up a handful of letters from Cincinnati supporters to leading politicians, Lowe traveled to Washington, where he enlisted Henry's support for the creation of a Union Army balloon corps.

Lowe would ultimately command a corps that included seven balloons, nine aeronauts, a series of wagons to generate hydrogen lifting gas in the field, and the world's first aircraft carrier, the *George Washington Parke Curtis*. The carrier, a 122-foot-long coal barge fitted with a flat deck and hydrogen-generating apparatus, enabled the members of Lowe's team to make observations up and down the tidal Potomac. In addition to active service in Northern Virginia, Lowe and his balloonists saw real action during the Peninsula campaign of 1862, at the Battles of Fredericksburg (1862) and Chancellorsville (1863), and with Federal forces on the Atlantic coast and the Mississippi River.

As a civilian, Lowe had great difficulty dealing with military bureaucracy and integrating his corps into the organization of the Union Army. Whenever possible, Henry interceded on his friend's behalf. He wrote to Secretary of War Edwin Stanton on July 21, 1863, noting that he had "much intercourse with Mr. Lowe" and had "formed a very favorable opinion of his skill and knowledge as an aeronaut, and of his trustworthiness as a man." He informed Stanton that Lowe "has had to contend with considerable prejudice, yet I doubt not that he has frequently obtained important information as to the position of the enemy for the use of which the officers alone were entirely responsible." At Lowe's request, Henry asked the secretary of war to assist Lowe "with the adjustment of his accounts."

Convinced that Army officials failed to appreciate his services, Lowe resigned in 1863, when his request for increased funding was refused. The Union balloons continued to be operated by members of Lowe's corps for a few more months but would never again see action. In 1864 Henry supported a request from Lowe to demonstrate his balloons before a federal commission that had been appointed to study new inventions and agreed that Lowe could once again ascend from the Smithsonian grounds. Government officials decided against giving the aeronaut a second chance, however.

SAMUEL P. LANGLEY MEDAL 1910
Presented by Charles Walcott to the Wright brothers, February 10, 1910

OPPOSITE: Crew members of the Union Army balloon corps use the balloon *Constitution* as a hydrogen gas reservoir to replenish the balloon *Intrepid*, May–June 1862.

Joseph Henry's interest in aeronautics in support of the Union cause was not limited to Lowe and his balloon corps. In the fall of 1861, he called the attention of Gen. George Brinton McClellan, commander of the Army of the Potomac, to the work of William H. Helme, a Rhode Island dentist who had experimented with aerial photography before the war. Helme suggested that hot air balloons, inflated by burning alcohol, would be easier to move, inflate, and manage than Lowe's gas balloons. When Lincoln's secretary, John Hay, also suggested that the general consider the idea, calling attention to Henry's commendation, McClellan instructed the War Department to provide Helme with $500 with which to build a demonstration craft. After a series of trials produced only mixed success, the idea was dropped.

On March 31, 1864, Henry and Maj. J. S. Woodruff of the Corps of Engineers met with Solomon Andrews, an ingenious fellow who had invented sewing machines, nutcrackers, tobacco filters, barrel-making machines, gas lamps, kitchen stoves, and a burglar-proof lock—all while serving three terms as the mayor of Perth Amboy, New Jersey. Since 1847, he had been struggling to develop an airship powered by gravity itself.

Following service as a surgeon with the Union Army, Andrews returned to

New Jersey in 1862 and began work on a full-scale *Aereon*, as he dubbed his flying machine. On June 1, 1863, he walked the *Aereon* out of its hangar and finished assembling the thing in the open. Three cigar-shaped balloons, each 80 feet long and tied together side by side, provided the lift. A large platform for the operator dangled beneath the gasbags, held in place by a confusion of ropes. Released with Andrews on the platform, the *Aereon* climbed at an angle as it moved forward, then wheeled around in the air and flew in the opposite direction, climbing to 200 feet before returning to a spot near the takeoff point for a safe landing.

How had he done it? The idea was almost as old as the balloon. At takeoff, the operator nosed the craft up by moving a heavy weight on a line to the rear of the platform. As the craft rose, the aerodynamic force on the overhead gasbags caused the craft to move obliquely forward. The operator used a large rudder to turn the craft, so that it was moving back toward the starting point as it continued to climb. To land, the operator moved the weight to the front of the platform, nosing the machine down. The inventor continued to give demonstration flights until September 4, when the *Aereon* was lost during an unmanned flight. The inventor reported that he had spent $10,000 of his own money on the craft.

Andrews now turned to Washington to fund an operational vehicle that could be used against the secessionists. He petitioned Congress for funding early in 1864, lectured in the capital on his scheme, and flew a four-foot model *Aereon* in the basement of the Capitol Building. Impressed, Robert Schenck, chairman of the Committee on Military

Constructed on the south side of the Smithsonian Castle in 1898, the South Shed housed Secretary Samuel Langley's aeronautical workshop. The 1903 Langley Aerodrome was constructed on the second floor of this building, which stood until 1975.

Affairs, asked Secretary of War Stanton to appoint a committee to study Andrews's claims and offer recommendations.

Two members of that committee, Secretary Joseph Henry and Maj. J. S. Woodruff of the Corps of Engineers, met with Andrews in the Smithsonian Castle on March 21, 1864. The third member of the group, Henry's friend Alexander Dallas Bache, Superintendent of the U.S. Coast Survey, was absent on business. After briefing Henry and Woodruff, Andrews flew his model in the library of the Institution.

"The inventor proved that the balloon can be steered," the Secretary remarked in a private note. "[It] can be made to move in an oblique direction while ascending or descending perpendicularly in still air. The power to stem a wind will depend upon the amount of ascensional power which the vehicle has to start with."

After prolonged discussions, the Secretary sent the committee's final report to Secretary Stanton in late July. Henry described Andrews as "one of the most ingenious and successful inventors of this country," and reported that the model *Aereon* could be steered in still air, and might even be made to move against a light wind. The group could not vouch for claims made for the larger craft, however, and recommended that Congress vote "a suitable appropriation" to fund a larger machine that "would fully test the question." But on March 22, 1865, the Committee on Military Affairs informed Andrews that his invention would not be required for the war effort.

JOHN STRINGFELLOW STEAM ENGINE 1868
The world's first aeronautical power plant

A dedicated researcher and manager of scientific research, Joseph Henry opposed, unsuccessfully, any effort to add the functions of a public museum to the Smithsonian's mission. His assistant, Spencer Fullerton Baird, who was appointed Secretary upon Henry's death in 1878, however, was dedicated to the creation of a national museum.

A graduate of Dickinson College in his native Pennsylvania, Baird went to work for Henry in 1850. His job was to free the Secretary of those functions he disliked, including the management and display of collections. A man of broad vision and interests, he used his position at the Smithsonian to foster science in the federal government. He played a key role in preparing many of the reports of U.S. government exploring expeditions, and in addition to his Smithsonian duties, he created and served as the head of the U.S. Commission of Fish and Fisheries, the first federal conservation agency.

Baird was also interested in public exhibitions as a means of educating Americans, and he took charge of the Smithsonian's participation in national and international fairs, notably the 1876 Centennial Exposition in Philadelphia. At the conclusion of the exposition, he made the rounds of both American and foreign exhibitors, persuading them to save the expense of shipping their exhibits home by donating them to the Smithsonian. Ultimately, he would ship 60 boxcars full of material to Washington, where it went into temporary storage in the Columbia Armory, the very place where T. S. C. Lowe had once prepared his balloon for flight. Two years later, Congress allocated the funds with which to build a new National Museum building, designed by architect Adolf Cluss and opened in 1881.

One-half of one of the halls of the new National Museum was devoted to the thousands of items donated by the Centennial Exposition's Chinese Imperial Commission. Buried somewhere in the 21 wagons required to transport the Chinese material from the train station to the Smithsonian were 42 hand-painted traditional festival kites. Constructed in the forms of men and women, children, folk characters, fish, insects, bats, flowers, and geometric designs, they were the first flying objects crafted by human hands to enter the Smithsonian collections. They remain today in the capable hands of curators at the National Air and Space Museum and the National Museum of Natural History.

THE AERODROME ERA

Feeling his age, Spencer Baird nominated a Smithsonian veteran, George Brown Goode, as assistant secretary for the National Museum in 1886, and a newcomer, Samuel Pierpont Langley, as assistant secretary for exchanges, publications, and the library. The Institution's governing Board of Regents confirmed the appointments on January 12, 1887. Less than a month later, on February 10, the Regents named Langley acting secretary in view of Baird's failing health. Upon Baird's death, Langley was appointed Secretary in his own right on November 18, 1888.

Samuel Langley combined Joseph Henry's abilities as a world-class scientific researcher with Spencer Baird's talent for the management of the scientific enterprise and determination to take full

Alexander Graham Bell, a longtime member of the Smithsonian's governing Board of Regents, shared the passion for aeronautics of his friend Samuel Langley.

TOP: In this 1907 photo taken by Washingtonian Carl Claudy, Bell stands in a field with several of his experimental tetrahedral cellular kites in the background.

CENTER: Washington photographer Carl Claudy catches a Bell tetrahedral kite in flight.

BOTTOM: One of Alexander Graham Bell's kites takes to the air.

advantage of the educational potential of the museum. A native of Roxbury, Massachusetts, he was the son of a prosperous produce merchant whose ancestors, including Richard, Increase, and Cotton Mather, had provided the intellectual leadership for the Puritan migration to New England in the 17th century.

A child of privilege, Sam Langley was educated in a series of private schools before entering the elite Boston Latin School, the nation's oldest school, at age 12. "As a child," a friend once wrote, "he was an omnivorous reader, had a reflective mind, an interest in art and in foreign lands, and a very strong bent toward mathematics." There was never any doubt as to where his real passion lay, however. "I cannot remember a time," he commented, "when I was not interested in astronomy."

Graduating from Boston High School in 1851, the ever practical Langley decided to follow his father into "the aristocracy of trade," rather than joining his Harvard-bound classmates. A career in architecture and civil engineering seemed more compatible with his interests and talents than the grocery business, however. The young man remained in the Boston area for the next six years, honing his drafting skills and establishing himself in his chosen professions. He struck out on his own in 1857, moving first to Chicago and then to St. Louis.

In 1864, after 14 years as an architectural draftsman and engineer, Langley steered his life in a radically new direction. He returned to Boston and in cooperation with his younger brother John—home from three years of Civil War service as a surgeon with the U.S. Navy—set to work on a large reflecting telescope. Their guide was an 1864 Smithsonian | *to page 42* |

When Samuel Langley's Aerodrome No. 5 made its first successful flight on May 6, 1896, the Secretary's close friend Alexander Graham Bell was on hand to witness the event. A member of the Smithsonian's governing Board of Regents, Bell launched his own pioneering aeronautical experiments following Langley's death in 1906.

ALEXANDER GRAHAM BELL

ON AERODROME NO. 5

From 1907 to 1909, Bell gathered a group of young enthusiasts into an organization called the Aerial Experiment Association. Bell's wife, Mabel, funded the development and testing of a series of flying machines, beginning with large man-carrying kites and concluding with such successful powered airplanes as the *June Bug,* in which Glenn Curtiss won the first Scientific American Trophy with a one-kilometer straight-line flight on July 4, 1908, and the *Silver Dart,* which made the first flight in Canada on February 23, 1909.

Bell's May 12, 1896, letter to the editor of the journal *Science* describes the peak moment in the aeronautical career of Samuel Langley.

. . .

Last Wednesday, May 6th [1896]. I witnessed a very remarkable experiment with Professor Langley's aerodrome on the Potomac River; indeed, it seemed to me that the experiment was of such historical importance that it should be made public.

I am not at liberty to give an account of all the details, but the main facts I have Professor Langley's consent for giving you, and they are as follows:

The aerodrome or "flying machine" in question was of steel, driven by a steam engine. It resembled an enormous bird, soaring in the air with extreme regularity in large curves, sweeping steadily upward in a spiral path, the spirals with a diameter of perhaps 100 yards, until it reached a height of about 100 feet in the air at the end of the course of about half a mile, when the steam gave out, the propellers which had moved it stopped, and then, to my further surprise, the whole, instead of tumbling down, settled as slowly and gracefully as it is possible for any bird to do, touched the water without any damage, and was immediately picked out and ready to be tried again.

A second trial was like the first, except that the machine went in a different direction, moving in one continuous gentle ascent as it swung around in circles like a great soaring bird. At one time it seemed to be in danger as its course carried it over a neighboring wooded promontory, but apprehension was immediately allayed as it passed 25 to 30 feet above the tops of the highest trees there, and ascending still further, its steam finally gave out again, and it settled into the waters of the river, not quite a quarter of a mile from the point at which it arose.

No one could have witnessed these experiments without being convinced that the practicality of mechanical flight had been demonstrated.

| from page 39 | publication in which New York astronomer Henry Draper described the process of producing a silver-coated glass telescope mirror and the other lenses, eyepieces, and mountings required for an instrument of professional quality. The Massachusetts instrument maker Alvan Clark was nearby to offer expert advice.

Samuel Langley was now in his element. He would earn his scientific reputation as an experimentalist, conceiving, designing, building, and operating precision instruments and mechanical devices, from bolometers capable of detecting minute changes in temperature over cosmic distances, to the jewel-like brass components of the small steam engines that would power his model aerodromes.

In 1865 the Langley brothers set off on a grand tour of European science and culture, visiting museums, historic sites, and observatories across the Continent. Upon his return to Boston, Samuel Langley learned that the Harvard College Observatory was expanding, and he introduced himself to Director Joseph Winlock. Impressed by the 30-year-old's enthusiasm and the quality of his work on the telescope, the astronomer hired Langley as an observatory assistant.

Anxious to make up for lost time, Langley accepted a position as assistant professor of mathematics at the U.S. Naval Academy in 1866, with the understanding that his primary duty would be to reestablish the school's observatory now that the academy had returned to Annapolis, Maryland, after temporary wartime relocation farther from the Confederacy in Newport, Rhode Island. He moved once again just a year later, accepting a post as professor of astronomy | to page 51 |

Alexander Graham Bell's kite experiments resulted in the design of Cygnet III, which left the ground in powered flight in March 1912 with John McCurdy at the controls.

1273

These Chinese kites were the first aeronautical objects to enter the Smithsonian Collections. The festival kite above, made of silk on a bamboo frame and decorated with a bear and snake motif, was originally shown at the Centennial Exposition in Philadelphia in 1876. The cicada kite opposite was one of a group presented to the Smithsonian by the Imperial Chinese Commission to the Philadelphia exposition.

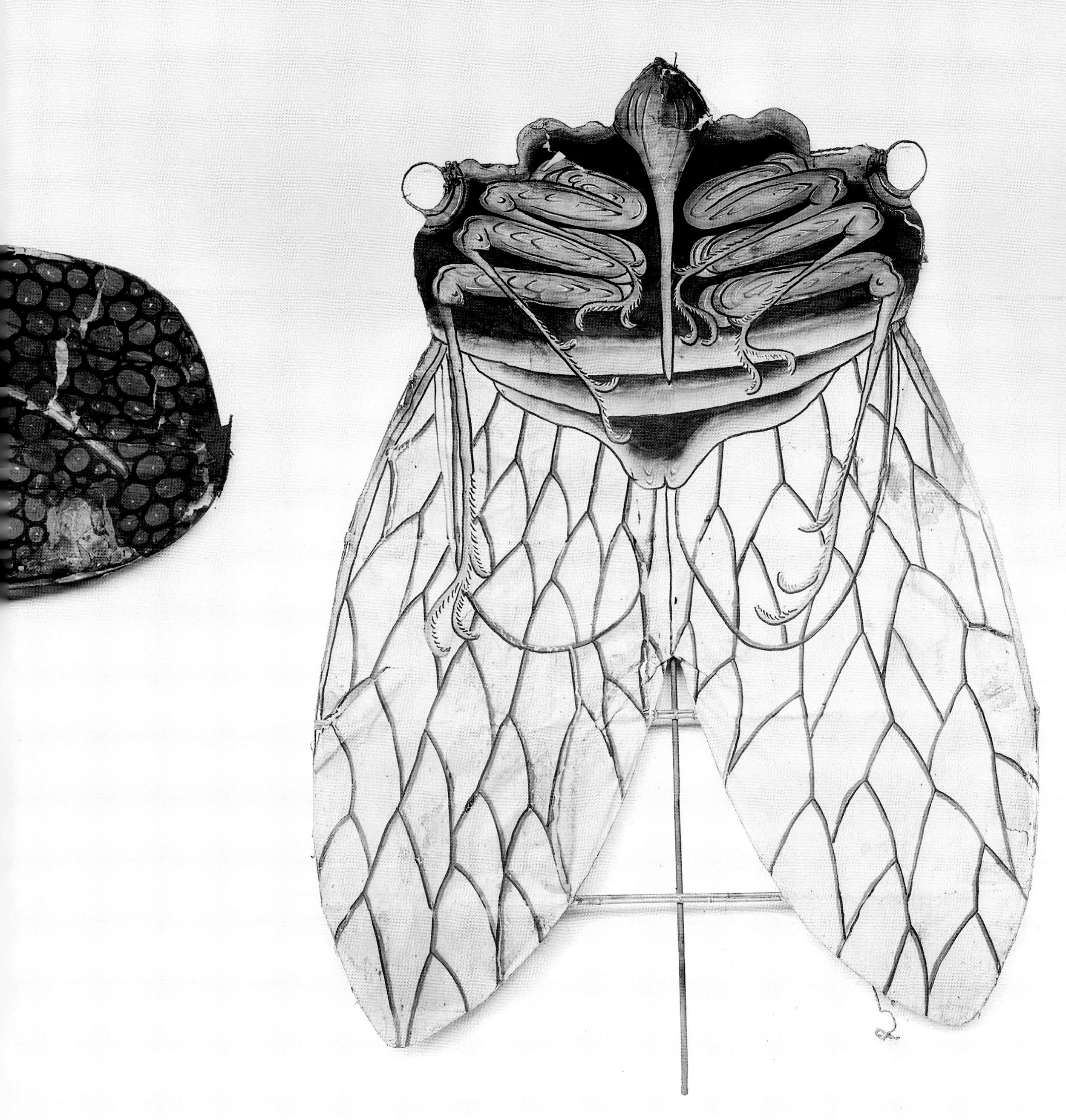

Acquired by the Smithsonian in 1906, this original 1896 Lilienthal glider is the only example of such a craft in the Western Hemisphere.

The German flight researcher Otto Lilienthal (1848–1896) was the single most important figure in the history of aeronautics between Sir George Cayley in the early 19th century and the Wright brothers at the beginning of the 20th.

LILIENTHAL GLIDER

A trained engineer, Lilienthal conducted a series of data-gathering experiments aimed at providing a foundation for wing design, the results of which he published in *Birdflight as the Basis of Aviation* (1889).

By 1890, assisted by his brother Gustav, he began to build and fly a series of gliders controlled by shifts in body weight. He flew from an artificial hill he constructed near his home in Berlin, and in the Rhinow Hills. By the time of his death—following a glider crash in August 1896—Lilienthal had accomplished as many as 2,000 glides. His experiences and recorded observations turned out to be significant contributions to increasing knowledge and understanding flight.

Lilienthal provided technical data of value to the succeeding generation of flight pioneers who would take the final steps to the invention of the airplane. In addition, the many photos of "the flying man" published in newspapers and magazines around the globe were convincing proof that the age of mechanical flight was close at hand.

The American newspaper magnate William Randolph Hearst purchased the Lilienthal Standardflugapparat—Standard Glider—from the German master in 1896. Hearst hired a local physical education instructor to make flights down the sloping lawn of his Long Island estate, but he chose to cease these demonstrations following Lilienthal's death.

Hearst gave the glider to aeronautical enthusiast John Brisbane Walker, who displayed it in a New York aeronautical exposition in 1906 and presented it to the Smithsonian a few months later.

The Lilienthal Standard Glider is the oldest successful piloted flying machine in the Smithsonian collection, and it is the only original machine built by Otto Lilienthal still to be found in the Western Hemisphere. It is now exhibited in the Early Flight gallery on the mall.

In this 1894 photo, Otto Lilienthal glides through the air in a craft similar to that on display in the National Air and Space Museum. Known as the "flying man," the German aeronautical pioneer died following a glider crash in August 1896.

Samuel P. Langley (in a light suit, with hands on lapel) and Alberto Santos-Dumont (with hands on hips) are among those attending a French aeronautical conference in September 1900.

| from page 42 | and physics at the Western University of Pennsylvania (now the University of Pittsburgh) and director of the university's Allegheny Observatory.

Arriving at his new post, he found that the observatory was equipped with an inadequate telescope. "Besides this," he wrote, "there was no apparatus whatever, not even a clock . . ." Before launching into his own research agenda, Langley would have to raise the funds with which to equip the observatory and support his work. In part, the answer was to be found in cultivating Pittsburgh's growing crop of industrialists, particularly William Thaw, whose investments in canals, railroads, and steamship lines made him one of the wealthiest citizens of one of the nation's wealthiest cities.

Beyond his connection to wealthy philanthropists, Langley put the observatory on a paying basis by selling time to the railroads. To coordinate train movements, maintain schedules, and prevent accidents, clocks and watches all along a railroad line need to be synchronized. Earlier, railroad officials had attempted to achieve this by moving an accurate clock from station to station. Langley proposed that accurate time, based on astronomical observations, be telegraphed instantly to every station on the line. This ingenious link between science and business put the Allegheny Observatory, and Samuel Pierpont Langley, on the national map.

Having made his reputation as an effective and imaginative manager of the scientific enterprise, Langley now proceeded to establish himself as a researcher. Pittsburgh, with its growing clouds of industrial smoke, was not the ideal location for | to page 54 |

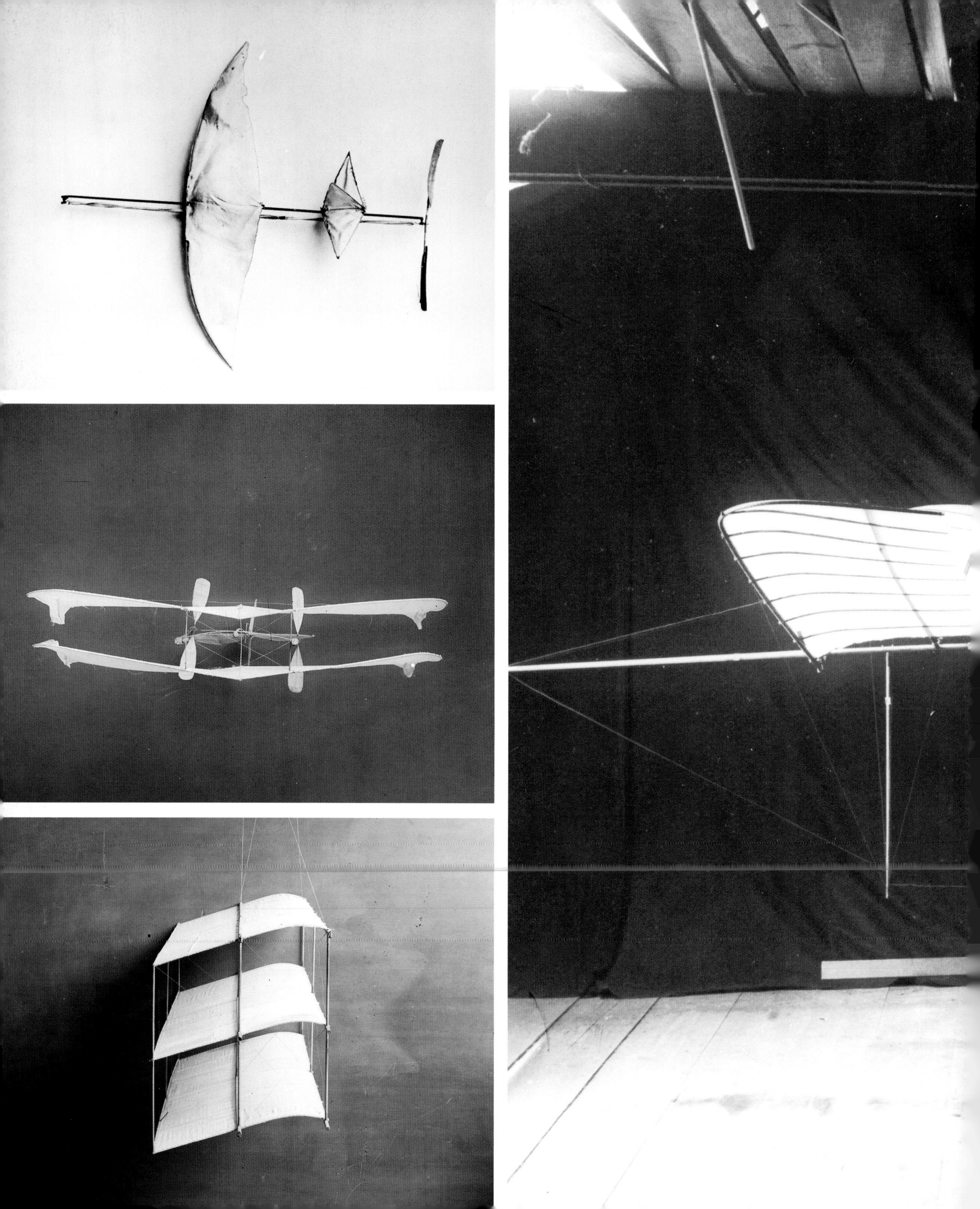

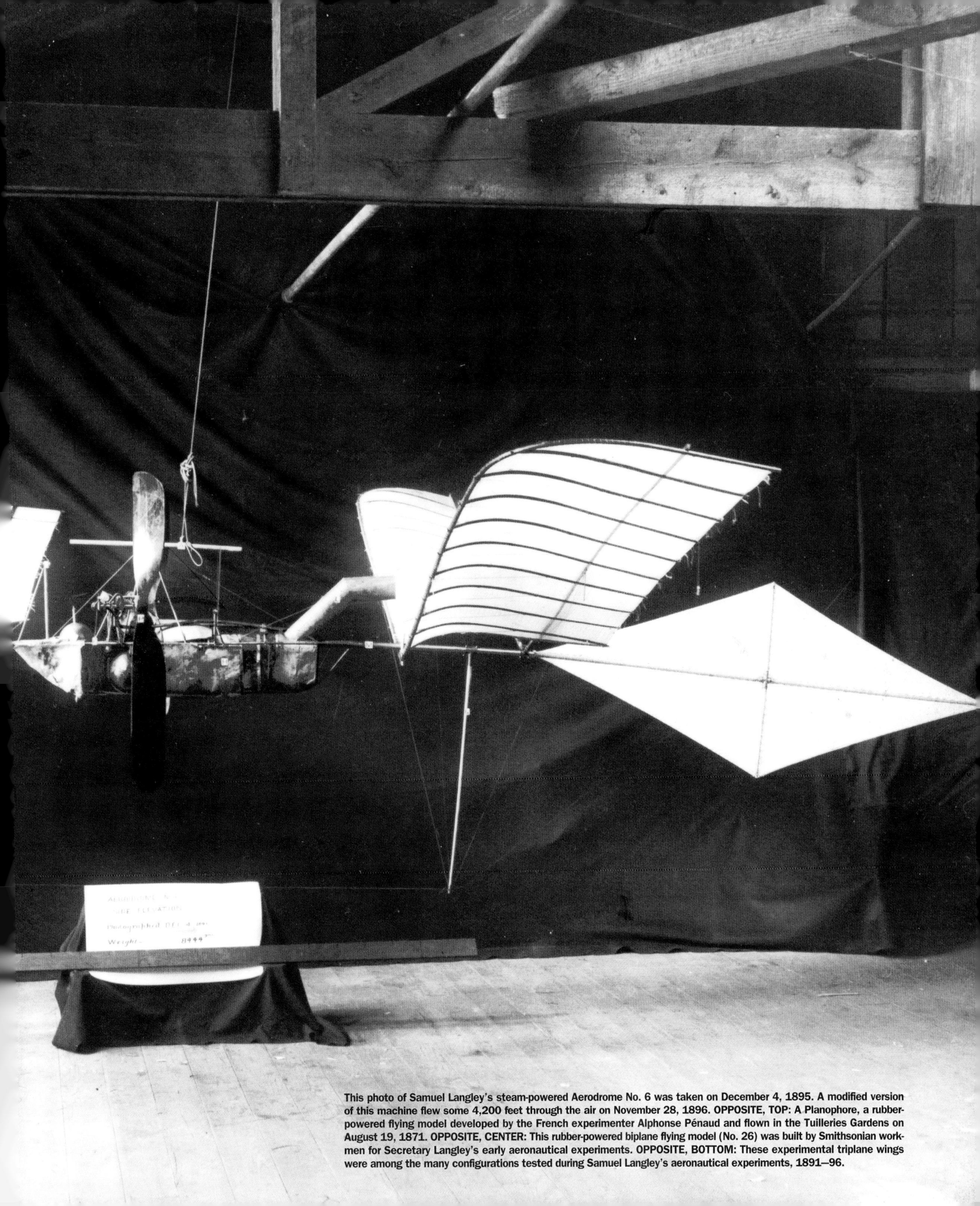

This photo of Samuel Langley's steam-powered Aerodrome No. 6 was taken on December 4, 1895. A modified version of this machine flew some 4,200 feet through the air on November 28, 1896. OPPOSITE, TOP: A Planophore, a rubber-powered flying model developed by the French experimenter Alphonse Pénaud and flown in the Tuilleries Gardens on August 19, 1871. OPPOSITE, CENTER: This rubber-powered biplane flying model (No. 26) was built by Smithsonian workmen for Secretary Langley's early aeronautical experiments. OPPOSITE, BOTTOM: These experimental triplane wings were among the many configurations tested during Samuel Langley's aeronautical experiments, 1891–96.

| *from page 51* | traditional optical astronomy. The sun, Langley noted, "is perhaps the only celestial object that I can hope to see and observe in any detail and with any regularity in the Pittsburgh area."

The astronomer began with visual observations of the solar surface, producing exquisite drawings of the dark spots appearing and disappearing on the face of the sun. He traveled to points in the U.S. and Europe where a solar eclipse would permit him a momentary view of the solar flares thrown far out into space. Noting that the sun was the celestial body that made all life on Earth possible, he turned from visual observations to the study of the solar energy reaching the Earth. Since the early 19th century, astronomers had known that the spectral lines produced by passing light through a prism exhibited varying temperatures. Moreover, it was clear that even higher temperatures were to be found in the areas of the spectrum above the visible light.

Langley, with his gift for instrumental work, designed the first bolometer, a device that could provide a precise measurement of the temperatures found in the spectrum. Recognizing that the total amount of solar energy reaching the surface of the Earth before it is absorbed by passage through the atmosphere (the solar constant) was key to understanding the impact of the sun on the Earth, Langley organized an expedition to California's Mount Whitney in 1881 so that he could take measurements above one-third of the atmosphere.

By the early 1880s, Samuel Langley was regarded as one of the leaders in what he and others were calling the new astronomy, or astrophysics. He was establishing a stellar scholarly reputation as an observer and experimenter, but he was just as interested in communicating his findings to the general public. In 1888, the year in which he formally became Secretary of the Smithsonian, Langley published *The New Astronomy,* a classic piece of popular science writing.

Fifty-two years old in 1886, Langley was a complex character. Physically imposing, he stood over six feet tall and weighed some 200 pounds. A fastidious and very formal dresser, he presented a formidable and, some thought, cold and stuffy persona to the world. His closest friends were quick to defend him as a shy and very private man.

The historian Henry Adams was among those who regarded Langley as "the great . . . master of experiment." At the same time, he was capable of surprising those who expected him to be entirely focused on science. A member of the St. Botolph Club, the home-away-from-home for Boston's literary and artistic elite, he was something of an aesthete. His best friend in Washington was not a scientist but a Smithsonian librarian and specialist in Judaica and folklore, Cyrus Adler. After attempting unsuccessfully to strike up a conversation with Langley on astronomy and physics, a Washington socialite finally asked what did interest him. "Children," he responded, "and fairy stories."

It seems likely that Spencer Baird's decision to hire Langley and the Board of Regents' decision to name him as the third Secretary of the Smithsonian were a result of his record as an effective fund-raiser and manager, his growing scientific reputation, and his gift for popularizing science. During his almost two decades at the helm of the Smithsonian, Langley would strengthen both the "increase and diffusion" sides of the Institution, establishing the National Zoo and the Smithsonian Astrophysical

OPPOSITE: In this 1933 painting artist Garnet Jex captured the moment when Samuel Langley's Aerodrome No. 5 took to the air for the first time, on May 6, 1896.

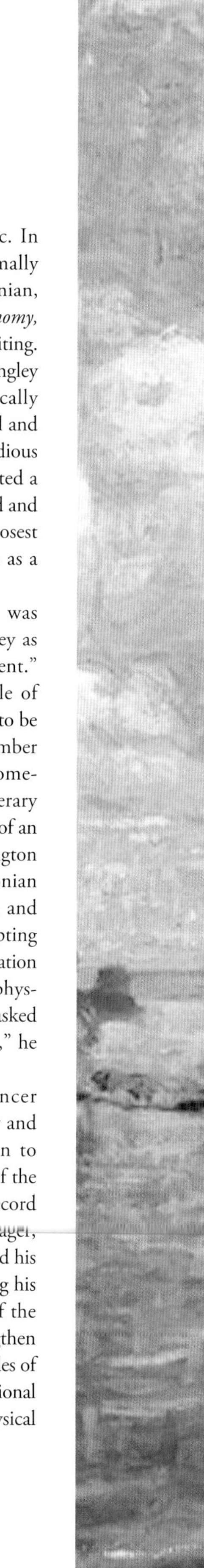

TACHOMETER 1887–1903 Instrument used by Samuel Langley to gauge engine performance

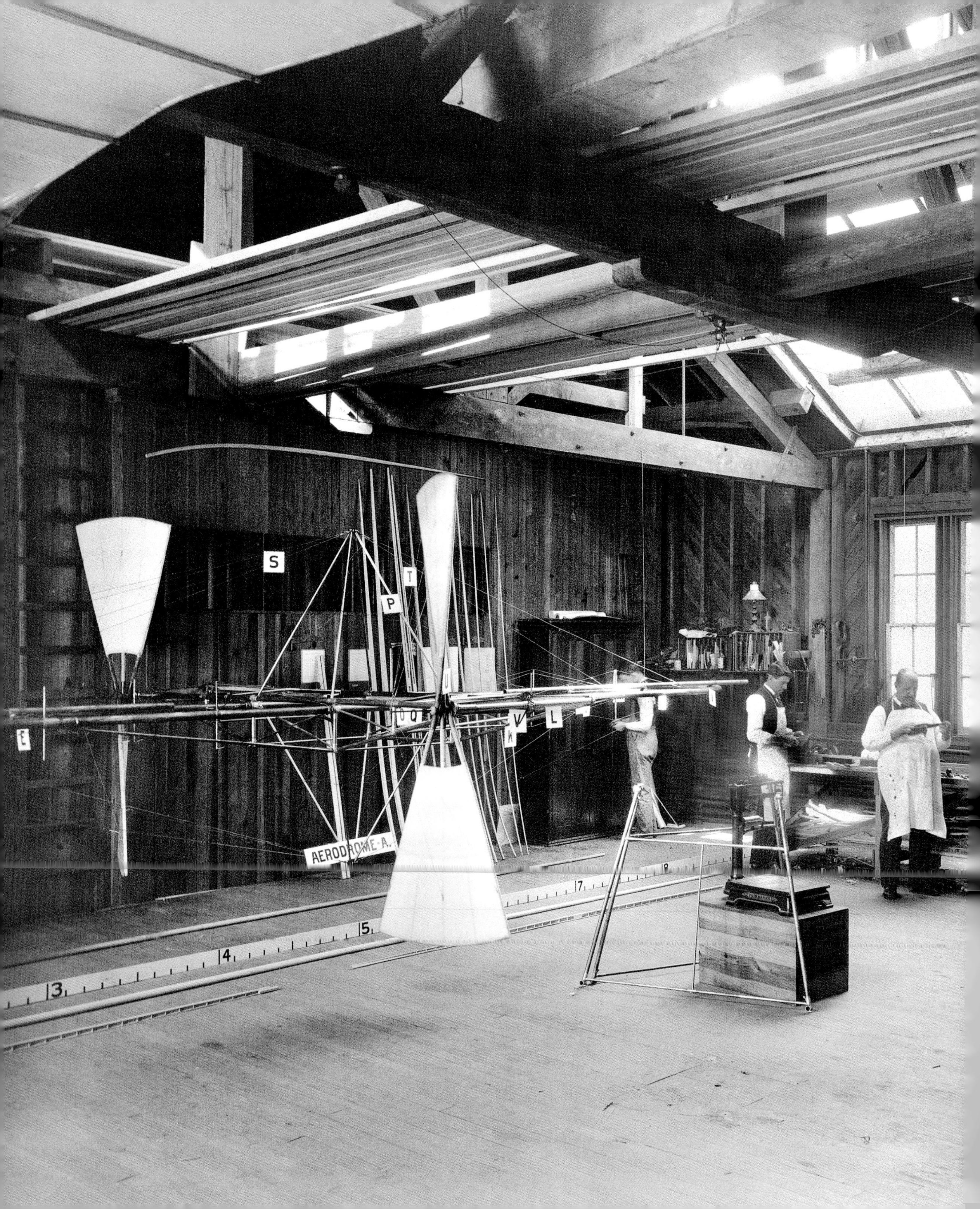
S
T
P
L
E
AERODROME-A.
3
4
5
7

The frame of the Great Aerodrome takes shape in the workshop on the second floor of the Smithsonian's South Shed in this photo taken on January 31, 1900.

Observatory; promoting the role of art at the Smithsonian, including opening negotiations leading to the establishment of the Freer Museum; and emphasizing the public education functions of the National Museum. One of the achievements of which he was most proud was the creation of a Children's Room in the south entrance to the Smithsonian Castle, where young museum visitors found appropriately sized display cases and labels designed to spark curiosity. "Knowledge," read the sign over the entrance, "begins in wonder."

Today, Samuel Langley is more often remembered for his failure to fly than for his undoubted achievements as Secretary. "The subject of flight interested me as long ago as I can remember," he wrote in 1897, "but it was a communication from Mr. Lancaster, read at the Buffalo meeting of the American Association for the Advancement of Science, in 1886, which aroused my then dormant attention to the subject."

Octave Chanute, one of the nation's leading civil engineers and a flying machine enthusiast, had arranged a session on aeronautics for the meeting in question. One of the presenters, Israel Lancaster, offered "The Soaring Bird," a paper that included a description of some flying models he had constructed. The session did not go well, and a local newspaper reported that the assembled scientists "unanimously joined in reviling and laughing [at]" Lancaster.

But Langley was not laughing. Intrigued, he sought to answer a basic question, which he later expressed as, "What amount of mechanical power was requisite to sustain a given weight in the air, and make it advance at a given speed?" Early in 1887 he set to work on a giant whirling arm, 60 feet from tip to tip, powered by a steam engine. The notion was to mount a variety of ingenious experimental devices on the end of the arm, then set it in motion, sweeping around in a great circle, gathering data on the reaction of solid surfaces in a fluid stream. The work was done at the Allegheny Observatory, where Langley was still spending part of his time, and funded by William Thaw.

As Langley began to spend more and more time in Washington after the fall of 1887, his experiments were gradually transferred to the capital, where he operated a smaller whirling arm on the Smithsonian grounds. The Secretary published the results of his work in 1891. "The most important general inference from these experiments," he concluded, "is that so far as the mere power to sustain heavy bodies in the air by mechanical flight goes, such mechanical flight is possible with engines we now possess." In fact, he argued—incorrectly, as it turned out—that the faster a body moved through the air, the less power would be required to sustain it.

Langley's aerodynamic work drew some criticism from conservative colleagues. Determined to demonstrate the accuracy of his data, the Secretary began, in April 1887, to work toward the construction of a mechanical flying machine based on his information and basic principles. He referred to his efforts to design and build such a craft as aerodromics, a term that he inaccurately derived from the Greek for "air runner." The flying machines themselves would be called aerodromes.

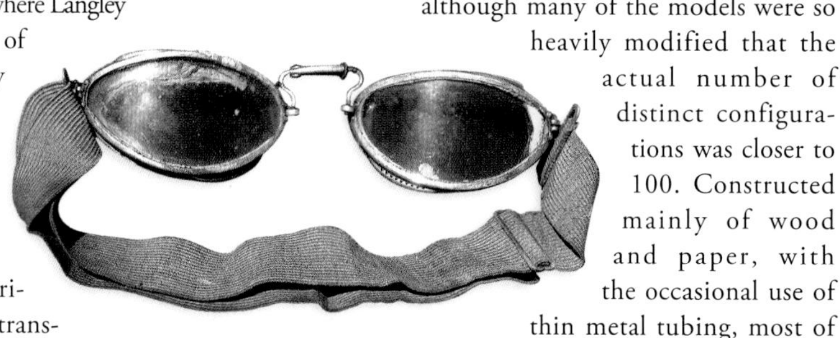

AVIATOR'S GOGGLES 1918 Issued to U.S. pilots during World War I

OPPOSITE: Charles Matthews Manly (left) and Samuel P. Langley pose on the roof of the houseboat on which Langley lived while working on the Great Aerodrome. This picture was taken soon before the craft's final December 8, 1903, test.

His starting point was the Planophore, a hand-launched, winged model airplane powered by twisted rubber string and developed by the French researcher Alphonse Pénaud in 1871. By 1891, Langley's workers in Pittsburgh and Washington had constructed 30 to 40 of the little craft, although many of the models were so heavily modified that the actual number of distinct configurations was closer to 100. Constructed mainly of wood and paper, with the occasional use of thin metal tubing, most of the models featured two propellers, mounted side by side or at the front and rear. There were monoplanes, biplanes, and craft with tandem wings. Most of the wings were flat, although some were curved, or cambered.

The little craft were flown in the Great Hall of the Smithsonian Castle. The Secretary quickly discovered, however, that the proportion of wing area to weight was much greater than in the case of birds. Only when the surface area of the wings approached four square feet could he hope for any sort of significant flight. No flight lasted longer than eight seconds or covered more than 100 feet.

But Langley was far from discouraged. His faith in his experimental data was so firm that he took a giant leap into the unknown, deciding to provide an incontrovertible demonstration of winged flight by building and testing much larger and heavier models with metal frames, propelled by lightweight steam engines. When the Secretary had experimented with hand-launched rubber-powered models, he needed little more than a carpenter to assist him. Now, however, he would have to create a workshop staffed by carpenters, metalworkers, and machinists. Funding would be drawn from grants that Langley had obtained for his astronomical work and from special Smithsonian accounts, such as the fund established at the Smithsonian by Thomas Hodgkins in 1891 to support research and publication on "the nature and properties of atmospheric air or for practical applications to the welfare of man."

Workers completed the first powered craft, Aerodrome No. 0, in the spring of 1892. Weighing in at 44 pounds, almost twice its predicted weight, it was abandoned without testing. Over the next year, the Smithsonian team began work on Aerodromes No. 1, 2, and 3, all of which were abandoned as being either too weak or too heavy and underpowered. Langley's method can be characterized as cut-and-try. Rather than working out the precise details of a machine before building it, he preferred that his workers actually build a model, then determine whether it was worth testing. As a result, they moved back and forth, searching for a balance between a structure that was too heavy and one that was too light and insubstantial to fly.

While his workers struggled to produce a suitable airframe and engine, Langley was considering the details of a test program. He decided that it would be safest to launch and retrieve the aerodromes from a large body of water. He acquired a small houseboat, just 30 feet long and 12 feet wide, in November 1892 and ordered that a two-room shack be constructed on the deck with a spring-powered catapult and a 20-foot-long launch rail mounted on the roof. When the time came, a tug would tow the | to page 62 |

Alexander Graham Bell photographed Langley's Aerodrome No. 5 as it flew some 3,300 feet over the Potomac River on May 6, 1896. OPPOSITE: Samuel Langley's dream of flight came to a disastrous end when the Great Aerodrome flipped onto its back and entered the water, December 8, 1903.

| from page 58 | houseboat 20 miles down the Potomac to Chopawamsic Island, near the isolated village of Widewater, now Quantico, Virginia.

The first unsuccessful attempts to launch an aerodrome (Number 4) into the air began in November 1893. During the following months, the Smithsonian workers discovered that the slightest wind would prevent them from even mounting the craft on the launcher. Once that was accomplished, they encountered one mechanical difficulty after another. Efforts were renewed in 1894 with no more success. Nor did things go any better during the 1895 season.

As the problems mounted, Langley decided to hire some experienced "aerodromic assistants." The Chicago engineer, Octave Chanute, introduced him to two such men. Langley hired Augustus Moore Herring, a Georgia native who in May 1894 had built and flown hang gliders, inspired by the work of German experimenter Otto Lilienthal. He brought Edward Chalmers Huffaker, an East Tennessee man and an engineering graduate of the University of Virginia, on board in December 1894. Both men made important contributions to the final success of the model aerodrome project, and both found Langley to be a difficult boss.

The Secretary may have loved nature, children, and fairy tales, but he was also, at his worst, an impatient, demanding perfectionist who insisted on maintaining absolute control over his employees. Herring resigned in a huff in November 1895. Huffaker remained on the staff until 1899. In fairness to Langley, both Huffaker and Herring were themselves difficult men. When Octave Chanute paid them to spend time in the Wright brothers' camp at Kitty Hawk, North Carolina (Huffaker in 1901, Herring in 1902), the Ohio brothers could not get along with either of them.

Testing resumed on May 6, 1896, with two new models, the product of a six-year effort. Aerodrome No. 5, constructed of steel tubing with a wooden, silk-covered tail, would fit, fully assembled, into a box 14 feet square and 4 feet tall. Weighing 25 pounds, it was powered by a gasoline-fired, one-cylinder steam engine producing one horsepower. Aerodrome No. 6 was virtually identical.

Langley arrived by train early that afternoon, accompanied by his friend Alexander Graham Bell, a Regent of the Smithsonian. Number 6 was damaged while being positioned on the launch rail. Number 5 was sent down the launcher at 3:05 p.m. Before the stunned observers, with a Smithsonian photographer doing his best to capture it all on film, the machine dropped toward the water, then climbed and began two large circles overhead, finally landing on the water astern of the boat, having covered a distance of 3,300 feet through the air at a speed of 20 to 25 miles an hour. Retrieved, dried, and launched a second time at 5:10 p.m. Number 5 climbed to an altitude of 60 feet and flew three large circles, covering 2,300 feet.

MECHANICAL DRAWING SET LATE 19TH CENTURY
Used for drafting by engineer and aviation pioneer Octave Chanute

It was an extraordinary afternoon. For the first time in history, a relatively large winged, heavier-than-air machine had taken to the air and flown a significant distance under its own power—twice. Willing to rest on his laurels for a time, Langley left on a European tour soon after those first flights, instructing the staff to have Aerodrome No. 6 ready for testing in the fall. Finally tested on Saturday, November 28, 1896, the craft flew a distance of 4,200 feet at a speed of 30 miles per hour.

The nation's newspapers had shown extraordinary patience, keeping up the public interest in the Smithsonian's aeronautical experiments during the long years when there was little real progress to report, and the temptation to ridicule the Secretary must have been very great. Now there was no limit to their praise. And it was well deserved. Langley had conducted a program of solid aerodynamic research, proving that mechanical flight was theoretically possible. Then he had risked his reputation to prove his point, achieving a goal that many thought was impossible. It was obvious to everyone that if Langley could get his aerodromes into the air and keep them there for significant distances, the age-old dream of aerial navigation would soon be realized.

A more cautious, less self-confident man might have walked away from the problem, content in the knowledge that he had provided the most impressive demonstration of the possibility of powered, heavier-than-air flight to date. Langley, however, could not resist the temptation to take one more step, building and flying a piloted version of his aerodromes. Thus far, the Secretary had funded his aerodromic work by drawing on internal Smithsonian resources.

But to forge ahead with the construction of a full-scale machine capable of carrying a pilot, he would require outside funding. "If anyone were to put at my disposal the considerable amount—fifty thousand dollars or more—for . . . an aerodrome carrying a man or men, with a capacity for some hours of flight," the Secretary wrote his friend Octave Chanute in June 1897, "I feel that I could build it and should enjoy the task." He underscored his query six months later: "If you hear of anyone who is disposed to give the means to such an unselfish end, I should be glad to meet him."

In October 1897 Langley drafted a memo outlining his plans for the full-scale machine, "in the event that I may be called upon officially to pursue these investigations for the Government." In the early spring of 1898 Langley's friend Charles Doolittle Walcott, director of the U.S. Geological Survey, raised the issue with John Addison Porter, President William McKinley's secretary; Assistant Secretary of War George Meiklejohn; and Assistant Secretary of the Navy Theodore Roosevelt, all of whom favored funding the project. Remarking that this is "well worth doing," Roosevelt was among those who approved Walcott's suggestion that a committee be appointed to study the project and offer recommendations.

That board—including Walcott, Alexander Graham Bell, and representatives of the Army and Navy—met at the Smithsonian in early April, deliberated for a week, and five days after war had been declared with Spain, recommended that $50,000 be provided to fund Langley's aerodrome project. On the basis of that recommendation, Langley was instructed to present his proposal to the Board of Ordnance and Fortification, an Army board charged with funding and managing the development of promising new weapons. The Secretary finally met with the board on November 6, 1898. After considerable negotiations to assure that he would have full responsibility for the project, with minimal military oversight, Langley accepted the first half of what would be a $50,000 grant. The Smithsonian was in the flying machine business.

Work on the Great Aerodrome, or Aerodrome A, as it was sometimes called, would be undertaken in the carpentry and machine shops located in the South Shed, a two-story brick-and-block building behind the Smithsonian Castle. As with the model aerodromes, Langley would depend on the skill of his own workers, with R. Luther Reed heading up the construction crew. The Secretary knew that he would need professional engineering assistance. Discouraged by his experience with the veteran aeronautical experimenters Herring and Huffaker, he decided to recruit an aerodromic assistant with no aeronautical background. He settled on Charles Matthews Manly, a graduating senior in mechanical engineering from Cornell University.

Langley was confident that he had only to scale up the 1896 design four times to enable it to carry a pilot. Of course, some changes would be required.

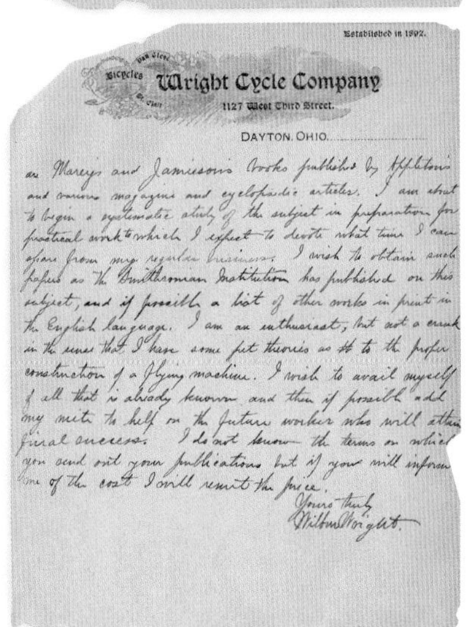

Wright, W. Established in 1892.

Bicycles — Van Cleve — St. Clair

Wright Cycle Company

1127 West Third Street.

DAYTON, OHIO. May 30 1899.

The Smithsonian Institution.
Washington;

Dear Sirs;

I have been interested in the problem of mechanical and human flight ever since as a boy I constructed a number of bats of various sizes after the style of Cayley's and Penaud's machines. My observations since have only convinced me more firmly that human flight is possible and practicable. It is only a question of knowledge and skill just as in all acrobatic feats. Birds are the most perfectly trained gymnasts in the world and are specially well fitted for their work, and it may be that man will never equal them, but no one who has watched a bird chasing an insect or another bird can doubt that feats are performed which require three or four times the effort required in ordinary flight. I believe that simple flight at least is possible to man and that the experiments and investigations of a large number of independent workers will result in the accumulation of information and knowledge and skill which will finally lead to accomplished flight.

The works on the subject to which I have had access

Established in 1892.

Bicycles — Van Cleve — St. Clair

Wright Cycle Company

1127 West Third Street.

DAYTON, OHIO.

are Marey's and Jamieson's books published by Appleton's and various magazine and cyclopaedic articles. I am about to begin a systematic study of the subject in preparation for practical work to which I expect to devote what time I can spare from my regular business. I wish to obtain such papers as the Smithsonian Institution has published on this subject, and if possible a list of other works in print in the English language. I am an enthusiast, but not a crank in the sense that I have some pet theories as to the proper construction of a flying machine. I wish to avail myself of all that is already known and then if possible add my mite to help on the future worker who will attain final success. I do not know the terms on which you send out your publications but if you will inform me of the cost I will remit the price.

Yours truly
Wilbur Wright.

LETTER
MAY 30, 1899
Wilbur Wright's letter to the Smithsonian announcing his interest in aeronautics

The lightweight steam-power plants would be replaced by a more powerful internal combustion engine. In November 1898, Richard Rathbun, assistant secretary, wrote to Stephen M. Balzer, a Hungarian-born, New York–based automobile builder, asking for a bid on a 12-horsepower engine weighing no more than 120 pounds, including the cooling system. Balzer proposed building a five-cylinder version of the rotary engine that he was currently manufacturing for automobiles. Ultimately, he would also build a one-quarter-size version of his engine to power a quarter-scale model of the Great Aerodrome.

In early June 1899, as aeronautical excitement and enthusiasm were building at the Smithsonian, the Secretary's office received a two-page letter on the pale blue letterhead of the Wright Cycle Company, at 1127 West Third Street in Dayton, Ohio. The correspondent, Wilbur Wright, announced that he had been interested in aeronautics since childhood and was convinced that "human flight is possible and practicable." Assuring Smithsonian officials that he was "an enthusiast, but not a crank," he requested copies of Smithsonian publications on mechanical flight and any suggestions as to other useful readings in English.

It was the sort of run-of-the-mill request that Richard Rathbun handled well. He responded | to page 69 |

The remains of the 1903 Langley Aerodrome float in the Potomac River following the first unsuccessful attempt to fly, October 7, 1903.

In 1892 Samuel P. Langley hired Cyrus Adler to serve as the librarian of the Institution. His hope was that the young scholar, a specialist in Judaica and folklore with a Ph.D. from Johns Hopkins University, would serve to offset the purely scientific focus of the Smithsonian staff. During Adler's 13-year tenure, the two men became the best of friends.

CYRUS ADLER
ON SAMUEL LANGLEY

Adler accompanied Langley to Europe in 1894, when the Secretary was awarded an honorary degree from Oxford University, and Langley offered his thoughts on mechanical flight to a meeting of the British Association for the Advancement of Science attended by some of the world's most distinguished physicists. Here Adler recalls the meeting.

. . .

The great event of the meeting [1894] of the British Association was a joint session of the physical and mathematical sections to discuss the possibility of the flight of a heavier-than-air machine, now called an aeroplane. Langley detailed his experiments with the whirling table, propounded his theories, and expressed his firm conviction in the possibility of the flight of such a machine.

When Langley finished reading his paper, Lord Kelvin, considered the great physicist and mathematician of his time, with the aid of a blackboard demonstrated the fallacy of Langley's arguments, and one after another of the great scholars present supported Kelvin's arguments.

The last address was delivered by Lord Rayleigh, who, while agreeing with the mathematicians that Langley's theory was unsound, wound up with the extremely astute and scientific remark that, nevertheless, if Langley succeeded in doing it, then he would be right. Less than three years after this date, Langley succeeded in flying a machine . . . on the Potomac River.

. . .

Nine years later, Adler was present for the final trial of the Great Aerodrome on December 8, 1903.

So on a very cold December afternoon in the same year, well toward dark, another launching was attempted, this time in the presence of General Wallace of the Army and chief of the Board of Ordnance and Fortification, the Secretary of the Board, and we had with us Francis S. Nash, a Navy Surgeon, in case anything happened.

The machine got off at dusk and, as before, went down into the river. [Pilot Charles] Manly, clad only in a union suit, stockings and light shoes, was fished out, and by the time he was brought to the houseboat, his clothing had frozen to his body and had actually to be cut away by the surgeon, so cold was the day. He was promptly given about half a tumbler full of whiskey to prevent him from catching cold.

And this young man, well brought up and with the utmost respect for the distinguished officers who were present, under the shock and probably the whiskey, delivered the most voluble series of curses that I have ever heard in my life.

| from page 63 | to Wilbur Wright on June 2, probably the day on which he received the letter, enclosing four reprints of articles on flying from Smithsonian annual reports and providing the requested list of additional publications available for purchase. It must have seemed like a small matter to Rathbun, but it was very important to the Wrights. Wilbur's letter to the Institution was the first announcement to the world that they were interested in the flying machine problem. They were anxious to survey the literature and access the state of flying machine studies. Rathbun's package of materials and advice on additional readings was, quite literally, their first step down the path toward the invention of the airplane. In responding to a request for information from two small businessmen in so prompt and thoughtful a fashion, Richard Rathbun may well have contributed more to the invention of the airplane than all of the work on the Aerodrome project.

By August 1900, after a long string of missed deadlines, disappointing tests, and a great many cash advances, it was clear to Manly and Langley that, while the engine showed real promise, Balzer had taken it as far as he could. Manly closed the contract and brought both engines back to the Smithsonian, where work would continue aboard a "floating machine shop" installed on a new and much larger houseboat moored at the Washington docks at the foot of Eighth Street. That effort would continue through April 1902, as Manly and his workers transformed Balzer's original rotary engine into a fixed radial, complete with water jackets and a radiator. In the end, the Manly-Balzer engine developed a full 52 horsepower for a fully fueled weight, including water and oil, of some 200 pounds.

It was an outstanding achievement, purchased at a considerable price. A significant portion of the expenditure on the aerodrome project went to power plant development. The attention paid to the propulsion system represented time, energy, and resources that were not going to the rest of the machine.

While work on the engine seemed to drag on with no end in sight, Langley and Manly wrestled with other issues. The pair considered basic changes to the 1896 configuration—including biplane wings—and rejected them in favor of sticking as close as possible to a scaled-up version of the originals. In building the structure of the Great Aerodrome, Langley's workers struggled with the old dilemma of the trade-off between weight and strength. In the end, these decisions were to prove critical to success—or failure.

They were also giving thought to the problem of control. As unpiloted flying models, the 1896 aerodromes had to have a significant degree of inherent stability. They were designed to recover automatically when their balance was upset. Dihedral, angling the wings up from the fuselage to the wingtip, provided balance in the lateral, or roll axis. A slight negative angle given to the cruciform tail served to balance the aircraft in pitch. The Smithsonian team provided Manly with a movable rudder (ineffectively mounted under the midpoint of the machine) and a measure of pitch control through the ability to raise or lower the tail or elevator. For the most part, however, they were hoping that the Great

OPPOSITE: Wilbur and Orville Wright assemble their second glider in August 1901, working in a wooden shed erected in July to serve as a workshop and to house the glider in bad weather.

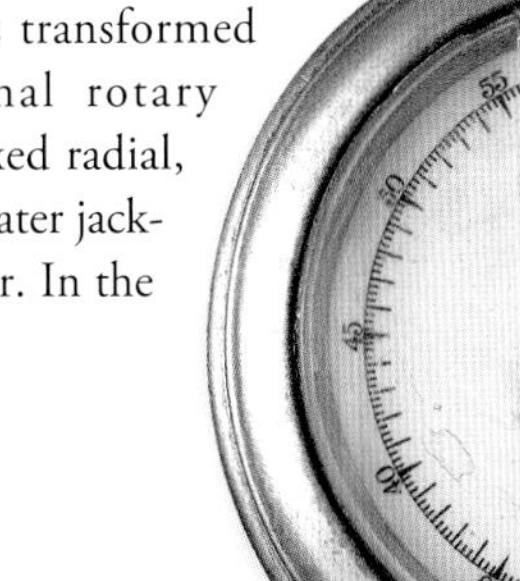

STOPWATCH CIRCA 1903 Used by the Wright brothers to time their four successful flights near Kitty Hawk, North Carolina, on December 17, 1903

Aerodrome would prove to be as inherently stable as its small predecessors. In effect, it was to be the biggest model airplane in the world.

As if to distract his team from an already overwhelming workload, Langley ordered the design and construction of a one-quarter-size model aerodrome built around a small 1.5-horsepower naphtha-fueled power plant provided by Stephen Balzer. Test launches of Numbers 5 and 6, as well as an unpowered one-eighth-scale model, were conducted beginning in the spring of 1899. Flight tests of the Quarter-Size Aerodrome followed on June 19, 1901. On the best of four flights, the little craft traveled only 350 feet. The engine and transmission were simply not up to providing sustained flight. When flown a second time in August 1903, the quarter-size model remained aloft for all of 27 seconds, covering 1,000 feet through the air. Newsmen who seemed determined to paint the Langley effort in the most positive light were more impressed than the performance of the Quarter-Size Aerodrome warranted. Still, it was the first heavier-than-air machine to make a successful flight powered by an internal combustion engine.

All of the pieces were finally coming together in the spring of 1903. When fully assembled, the Great Aerodrome measured 58 feet 5 inches from the tip of its bowsprit to the back of its tail, and 48 feet 5 inches from wingtip to wingtip. Without the wings and tail in place, the open fuselage, with its steel tube framing, coils of copper tubing, and shiny fixtures and fittings, looked more like a huge scientific instrument than a flying machine. With the large fabric wings attached, set at a rakishly high angle of attack, it could be described as a clipper ship of the clouds.

Seventeen years of effort stood behind the Great Aerodrome mounted on its new launcher aboard a new and much larger houseboat anchored at the familiar Potomac launch site at Widewater, Virginia, on October 7, 1903. With C. M. Manly seated in the fabric covered "cockpit," the craft shot down the launch rail at 12:15 p.m. "For the fraction of a minute the big airship soared," or, as Langley's hometown paper, the *Boston Daily Globe,* reported, "more properly speaking trembled, then, as if water and not air was its natural element, it headed downward and a few seconds later the machine was floundering in the Potomac, a hopeless wreck."

More direct and honest, a Chicago paper simply headlined: "Langley Airship a Total Failure." A reporter for the *Washington Post* supplied the graphic image that would stick in the public's mind when he noted that the Aerodrome had dropped into the river "like a handful of mortar." Langley had been too nervous to take the train down to the launch site that morning. Less fortunate, or farsighted, Manly was rescued without injury.

The Great Aerodrome did not escape so lightly. The steel tube frame survived, but the wings, tail, and propellers were smashed beyond repair. Langley announced that the accident had been the result of a failure of the catapult, and he promised that the machine would be ready for a second trial in a month.

By December 8, winter had arrived in the nation's capital. With chunks of ice floating in the Potomac, Langley decided that it would be foolish to travel back downriver. The second trial would be made with the launch boat and its cargo anchored in the Potomac, off Arsenal Point. It was late afternoon, almost dusk, when the aerodrome started down the track. A photograph taken before the machine reached the end of the launcher indicates that the rear wings were already folding up. With no lift at the rear, the craft nosed straight up and flipped over on its back, entering the water only 20 feet from the boat. Fred Hewitt, a Smithsonian mechanic, plunged into the freezing river and pulled Manly out from under the wreckage. Moments later, wrapped in blankets and fortified with whiskey, Charles Manly, the cultured son of a university president, startled the group gathered around him on the scow by delivering "the most voluble series of curses."

Nine days later, at the foot of a windswept dune on the Outer Banks of North Carolina, Wilbur and Orville Wright made the first powered, controlled, sustained heavier-than-air flights in history. Octave Chanute, a friend of both the Wrights and Langley, urged the Secretary to travel to North Carolina to meet with the brothers. Langley sent a telegram of inquiry to the little Weather Bureau telegraph at Kitty Hawk, but it arrived after the brothers had left for Dayton. While the success of the Wrights was widely, if inaccurately, reported in American newspapers over the next few days, it was generally overshadowed by the continuing discussion of Langley's failure to fly. | to page 76 |

OPPOSITE: Orville Wright returned to the Outer Banks of North Carolina in 1911 to test a new glider. On October 24 he remained aloft for 9 minutes 45 seconds, setting a world record for soaring flight that would stand for ten years.

FUSEFORM MODELS 1901 Aerodynamic shapes tested in a wind tunnel by Albert Zahm at Catholic University in Washington, D.C.

The Smithsonian's first full-scale successful powered airplane arrived in 1911, although not without a good deal of trouble. The Wright brothers had sold their 1909 Military Flyer, the world's first military airplane, to the U.S. Army.

1909 WRIGHT MILITARY FLYER

WORLD'S FIRST MILITARY AIRPLANE

This airplane was the historic machine that Orville Wright had demonstrated in spring 1909 at Fort Myer, Virginia, and that Wilbur Wright had used to teach the Army's first three aviators—Lt. Frank P. Lahm, Lt. Frederick Humphries, and Lt. Benjamin D. Foulois—to fly at College Park, Maryland, that fall. The Army then ordered Foulois to proceed with the airplane to Fort Sam Houston in San Antonio, Texas, to explore its military potential.

After two years and a good many mishaps and alterations, the Army shipped the machine back to the Wright company in Dayton with a request to put it back in "first-class condition" so that it could be used for training new aviators. Factory manager Frank Russell responded that refurbishing the airplane for additional service would be impractical and suggested that the Army consider donating it to the Smithsonian. When Gen. James Allen, the chief signal officer, offered

The 1909 Wright Military Flyer, the world's first military airplane, has a place of honor in the National Air and Space Museum's Early Flight gallery.

it in May 1911, Assistant Secretary Richard Rathbun opposed accepting the Flyer, arguing that space constraints made it "out of the question to hope for a comprehensive exhibit of actual aeroplanes."

Fortunately for the future of aeronautics at the Smithsonian, Secretary Walcott decided that the world's first military airplane was, after all, an object worthy of display in the National Museum and ordered it suspended in the West Hall of what was now called the Arts and Industries Building, after the opening of the Natural History Building across the mall in 1910. It remained there until the beginning of World War II, when it was moved to safe storage at a National Park Service facility in Luray, Virginia.

Today the 1909 Wright Flyer is the prize artifact in the National Air and Space Museum's Early Flight gallery.

The 1909 Wright Military Flyer being prepared for flight at Fort Myer, Virginia: Six soldiers haul on the launch cable as Orville Wright stands between the forward supports of the aircraft.

| *from page 70* | The Secretary announced that the second failure had once again been the result of a catapult malfunction. He had every intention of trying a third time. But the Great Aerodrome was more severely damaged than in October. More important, public opinion had turned against him at last. After calling attention to "the repeated and disastrous failures" of the Great Aerodrome, "the absurd atmosphere of secrecy in which it was enveloped," and the "expensive pageantry that attended its various manifestations," the editors of the *Washington Post* argued that the time had come for "a withdrawal by the government from all further participation in its financial and scientific calamities." Indeed, Army observers who were present for the December attempt reported that the craft was "not planned correctly for flight."

Within a week, Langley's failure was inspiring would-be humorists across the nation. The *Chicago Inter-Ocean* noted that while the aerodrome was not much of a flying machine, it might "rival the best of the submarine divers." The *Newark News* suggested that "a little dynamite" placed under the machine "might get more flying out of it than the launching apparatus hitherto used so disastrously." An Indianapolis wit commented that the aerodrome "is an aquatic bird, and will not be satisfied in any other element."

Cruel humor gave way to direct attacks early the following month. During consideration of an Army appropriations bill before the House Committee of the Whole on January 23, Representative James M. Robinson of Indiana attacked the secretary of war for approving the expenditure of "this money that was sunk in the bottom of the Potomac River by this aerial navigation, this Don Quixote scheme of Prof. Langley." The Secretary was described as a professor "wandering in his dreams of flight . . . given to building castles in the air." Representative Gilbert Hitchcock of Nebraska asked James A. Hemenway, chair of the House Appropriations Committee, "if it is to cost us $73,000 to construct a mud duck that will not fly 50 feet, what is it going to cost to construct a real flying machine?"

In mid-March, the Board of Ordnance and Fortification refused additional funding for the aerodrome project, "largely in view of the adverse opinions expressed in Congress and elsewhere," as the organization noted in its annual report. Even those who were most sympathetic to Langley were now finding fault with his management of the flying machine project. "There is," the *New York Times* noted, "not a little criticism of Prof. Langley in official circles for leaving the work of making and trying the Aerodrome to subordinates and not himself applying his remarkable scientific mind directly to the problem which he so confidently predicts he can solve."

Langley initially hoped that he could raise private funds with which to continue the work, but soon changed his mind. "If the government no longer wants it," he wrote to those who offered to assist in fund-raising for the project, "then I no longer feel obliged—nor do I desire—to pursue it." The Secretary ordered the frame of the aerodrome to be repaired and stored, along with the engine, transmission, and surviving elements of the machine. The physical remains of the aerodrome project, 1891–1903, were locked away on the second floor of the South Shed.

Secretary Langley suffered an even more staggering blow in June 1905, when auditors discovered that Smithsonian accountant W. W. Karr, a man whom Langley trusted implicitly, had defrauded the Institution of $68,558.61. Karr pleaded guilty and was sentenced to five years in the federal prison at Moundsville, West Virginia. Devastated, the Secretary would never again accept his Smithsonian salary. Langley suffered a stroke five months later. Showing signs of recovery, he traveled to Aiken, South Carolina, in early February 1906, accompanied by his niece (his brother John's daughter) and a doctor. He died of a second stroke suffered on February 27, 1906.

OPPOSITE, TOP: Charles Doolittle Walcott (1850–1927) served as the fourth Secretary of the Smithsonian Institution (from 1907 to 1927) and was instrumental in establishing the National Advisory Committee for Aeronautics.

OPPOSITE, BOTTOM: A view of a January 1906 exhibition in New York City sponsored by the Automobile and Aero Clubs of America. Langley's Aerodrome No. 5 is suspended in the foreground.

CURTISS V-8 MOTORCYCLE 1907
In January 1907 Glenn Curtiss rode this motorcycle at a world-record speed of 136 mph.

WALCOTT AND THE FUTURE OF FLIGHT

The Smithsonian Board of Regents named Charles Doolittle Walcott the fourth Secretary of the Institution on January 31, 1907. Like his close friend Samuel Langley, Walcott had little formal training in science. He was born in upstate New York and left school without a diploma. While working as a farm laborer, he became interested in fossils, and he assisted in building several collections that were purchased by universities. His experience in the field led in 1876 to his appointment as assistant to James Hall, New York State paleontologist. Three years later, he became one of the original members of the U.S. Geological Survey, a position that enabled him to explore the rich fossil deposits of the American West.

Walcott became the third director of the Survey in 1894.

When his old colleague George Brown Goode died in the fall of 1896, Secretary Langley asked Walcott to accept the post of assistant secretary, with responsibility for the National Museum, where the geologist had maintained an office for several years. He accepted, but only if he could remain as director of the USGS. Even as a part-timer, he had a major impact on the public side of the Smithsonian. Walcott finally gave up his post with the USGS after his appointment as Secretary of the Smithsonian.

While his own research interests related to the early history of life buried deep in the earth, he thought it important to build on his predecessor's initiatives in astronomy and aeronautics. In 1895 Langley hired Charles Greeley Abbot, an MIT graduate in chemical physics, as his astrophysical assistant, promoting him to acting director of the Smithsonian Astrophysical Observatory just a year later. Recognizing Abbot's extraordinary talents, Walcott named him director of the observatory in 1907. Although the new Secretary had no intention of continuing the aerodrome program, he was convinced the Institution should play a role in the development of American aviation.

By 1913, it was clear that leadership in aeronautics had passed from the United States, the birthplace of the airplane, to Europe, and the gap was growing wider. That year a congressional study revealed that the United States ranked 14th among the nations of the world in terms of government expenditures on aeronautics. Belgium, Japan, Chile, Greece, Bulgaria, Spain, and Brazil all outspent | *to page 81* |

Wilbur Wright flies over hay fields near Pau, France, with a passenger aboard—Capt. Paul N. Lucas-Girardville—in the early spring of 1909.

Carnation
Milk
BENOIST

| from page 77 | the United States. In a separate report, the secretary of the Navy noted that Japan spent $600,000 on aircraft and flight research in 1912, compared with $140,000 expended by the U.S. government.

A significant portion of the money spent on aeronautics by the English, French, Germans, and Russians went to support flight research in universities and in the national aeronautical laboratories being operated by each of those nations. Such research was aimed at solving problems common to aircraft or engine manufacturers and was an important means of strengthening national industry. One key to regaining a position of leadership in aeronautics for the United States was to establish an American aeronautical research organization.

As early as April 10, 1911, the *Washington Star* reported that plans were afoot to create a flight laboratory at the National Bureau of Standards under the auspices of the Smithsonian. The specter of interagency rivalry immediately reared its head when Rear Adm. A. M. Watt complained that such a plan would duplicate the existing research facilities at the Washington Navy Yard, which were fully equipped to investigate "a considerable portion of the phenomena associated with aeronautics." Richard Maclaurin, President of the Massachusetts Institute of Technology, agreed with Admiral Watt on the need for a national aeronautical research establishment, but he argued that it should be centered at a university, such as MIT, which could draw on the technical expertise of men from a wide range of fields.

Undaunted, Capt. Washington Irving Chambers, responsible for aviation matters in the Navy's Bureau of Navigation, enlisted the support of Albert Francis Zahm (1862–1954), a veteran aeronautical researcher who was teaching mechanical engineering and operating what was at the time the world's largest wind tunnel, at Catholic University. Together, the two began to make the case for an aeronautical laboratory at the Smithsonian. Zahm published a flurry of papers on the subject, and in September 1912, Chambers issued a well-organized 15-page memorandum outlining his plan for such a facility, operating under the supervision of a national board of aeronautical experts. The Smithsonian, he argued, could house such a small organization in the old Langley laboratory and workshop and could provide the political cover and organizational structure, as well as an atmosphere that would help to nurture such a new research initiative.

On December 16, 1912, President William Howard Taft appointed 19 members of a National Aerodynamical Laboratory Commission to study the question. Known as the Woodward Commission, in honor of its chairman, Robert Woodward, president of the Carnegie Institution of Washington, the group drafted a report accepting most of the points outlined in Chambers's memorandum. The question of whether the laboratory would be governed by the Navy, the Bureau of Standards, a university, or the Smithsonian remained contentious. Legislation based on the recommendation that the laboratory be created under the auspices of the Smithsonian failed to reach a vote, as did a separate attempt to appropriate funds for aeronautical research at the Institution.

OPPOSITE, TOP: Pilot William D. Jones took this extraordinary photo of himself while flying his Benoist Type XIV in 1913 over West Duluth, Minnesota. The camera is mounted on the outer wing. The pilot holds the shutter release in his hand.

OPPOSITE, BOTTOM: A. Roy Knabenshue, an airship pioneer and manager of the Wright brothers' exhibition flying team, treats his father, mother, and wife to a ride over Chicago aboard his airship in July 1914.

GORDON BENNETT BALLOON RACE PROGRAM 1906 Americans Frank Lahm and Henry Hersey won this first running of the Gordon Bennett classic.

Anxious to honor his predecessor's memory and solidify the Smithsonian's leadership in aeronautics, Secretary Walcott took matters into his own hands. On May 1, 1913, the Board of Regents voted to reopen what would now be known as the Langley Memorial Aerodynamical Laboratory. The new organization was to "plan for such theoretical and experimental investigations, tests, and reports as may serve to increase the safety and effectiveness of aerial locomotion for the purposes of commerce, national defense, and the welfare of man." The laboratory would operate under the guidance of an advisory committee that would include representatives of the Departments of War, Navy, Commerce and Agriculture, "and such persons . . . as may be acquainted with the needs of aeronautics." Subcommittees, each chaired by a member of the main committee but including nonmember specialists, would supervise work on individual projects or specific areas of research.

In the spirit of Samuel Langley, and as an apparent dig at the Wright brothers, who had brought suit for patent infringement against their rivals in the United States and Europe, Walcott stipulated that the new organization would not "promote patented devices, furnish capital to inventors, or manu facture commercially." On May 9, 1913, President Woodrow Wilson approved the designation of representatives to the new board.

The Advisory Committee to the Langley Aerodynamical Laboratory met in May, June, and December of 1913. The group had 11 members, 7 of whom were drawn from government agencies. There were 16 subcommittees, covering topics from applied aerodynamics to the dissemination of aeronautical information.

Other than Walcott, the key member of the committee was Albert Francis Zahm. Having cooperated with Chambers to lobby for a national research program, it was only natural that he be asked to serve as secretary of President Taft's study commission and as a member and recorder (secretary) of the final advisory board. For all of his contributions to the program, he was openly antagonistic toward the Wright brothers and was the central figure in initiating a long-running feud between the Smithsonian and Orville Wright over the flight capability of Langley's 1903 craft, the Great Aerodrome (see Interchapter 1.5, "Capable of Flight? The Wright-Smithsonian Controversy," pages 94–101).

The governing committee spent most of its first year surveying aeronautical specialists across the nation regarding those problems and research areas where the laboratory could have the greatest impact. The step that would have the farthest reaching consequences was to dispatch Zahm and Jerome Hunsaker, a young Annapolis graduate who was organizing an aeronautical engineering program at MIT, on a tour of European aeronautical laboratories.

The pair boarded the North German Lloyd liner *Kronprinz Wilhelm* on July 29, 1913. They traveled through France, Germany, and England, visiting every important aeronautical laboratory and research facility on the continent and speaking to the engineering leaders who had created those programs. Back in the United States that fall, they were determined to share what they had | *to page 86* |

This airman came to grief during San Francisco's 1915 Panama-Pacific Exposition.

TIO
TIO

53.21

An Albatross D.III and a Nieuport 17 face each other on the ground, as they did in the sky.

| from page 82 | learned and to push for a similar U.S. program.

It soon became clear, however, that such a program would not take root at the Smithsonian. In December 1913 Charles Walcott discovered that it was illegal for government officials to serve on an advisory board such as that governing the Langley Aerodynamical Laboratory without congressional approval. The Secretary immediately took the matter to the Smithsonian Board of Regents, which instructed him to request a congressional appropriation of $50,000 to support the work of the laboratory. Such an appropriation would not only provide additional support for an initiative that Walcott was underwriting with Hodgkins Funds, but would serve as de facto congressional approval for the program and the advisory committee.

Things did not go well when the Secretary took his proposal to the House Committee on Appropriations in March 1914. The prospect of catching up with European aeronautical rivals was not sufficiently appealing to justify funding a small initiative that might grow into an expensive new bureaucracy. To his credit, Walcott took a step back from the Smithsonian-centered laboratory and drafted a memo laying out the critical importance of aeronautical research to the nation. He explained to congressional leaders the value of the advisory committee, subcommittee structure, which was loosely based on the British Advisory Committee for Aeronautics and had been successful in coordinating flight research in that nation.

This time logic and good sense prevailed. Walcott's memo became the basis for legislation introduced in both the House and the Senate. A qualified endorsement from Assistant Secretary of the Navy Franklin D. Roosevelt, and Walcott's testimony before the House Naval Affairs Committee, removed any remaining concerns, but the session ended before the bill could be passed. It was reintroduced as an amendment to the naval appropriations bill, and President Wilson signed the legislation creating what would soon be known as the National Advisory Committee for Aeronautics (NACA) on March 3, 1915.

Launched with an annual budget of only $5,000, the NACA did not initially have a laboratory in which to conduct research. With Walcott as the head of the organization's executive committee, however, the NACA began to demonstrate its ability to coordinate the efforts of the growing community of aeronautical research organizations in the United States. Work on a real flight laboratory, the Langley Memorial Aeronautical Laboratory, to be located in Hampton, Virginia, began in 1917. Three years later Charles Doolittle Walcott became chairman of the NACA. He would hold that office, along with his Smithsonian position, until his death in 1927, continuing to shape the new agency into the key to U.S. success in the air that it would become.

By 1920, the years during which the Smithsonian Institution had played an important role in shaping the future of U.S. aeronautics were coming to an end. As the number of historic aircraft housed in the Aeronautical Annex—the "tin shed" on Independence Avenue behind the Smithsonian Castle—began to grow, the focus shifted from determining the future of flight to chronicling its history. That year a young man named Paul Edward Garber began work at the Institution. With his arrival, the history of the National Air and Space Museum was soon under way.

OPPOSITE: Aerial photography proved an invaluable resource to commanders during World War I. This 1918 Italian photo image shows the Austrian trenches near Monte San Michele in northern Italy.

DECORATED SHELL CASING 1917
An example of "trench art"

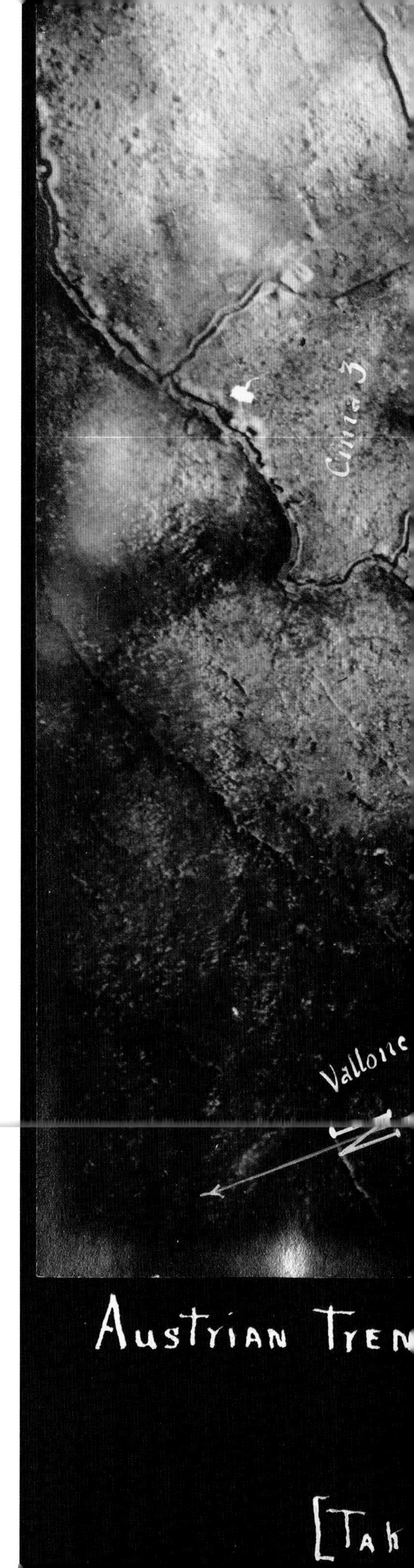

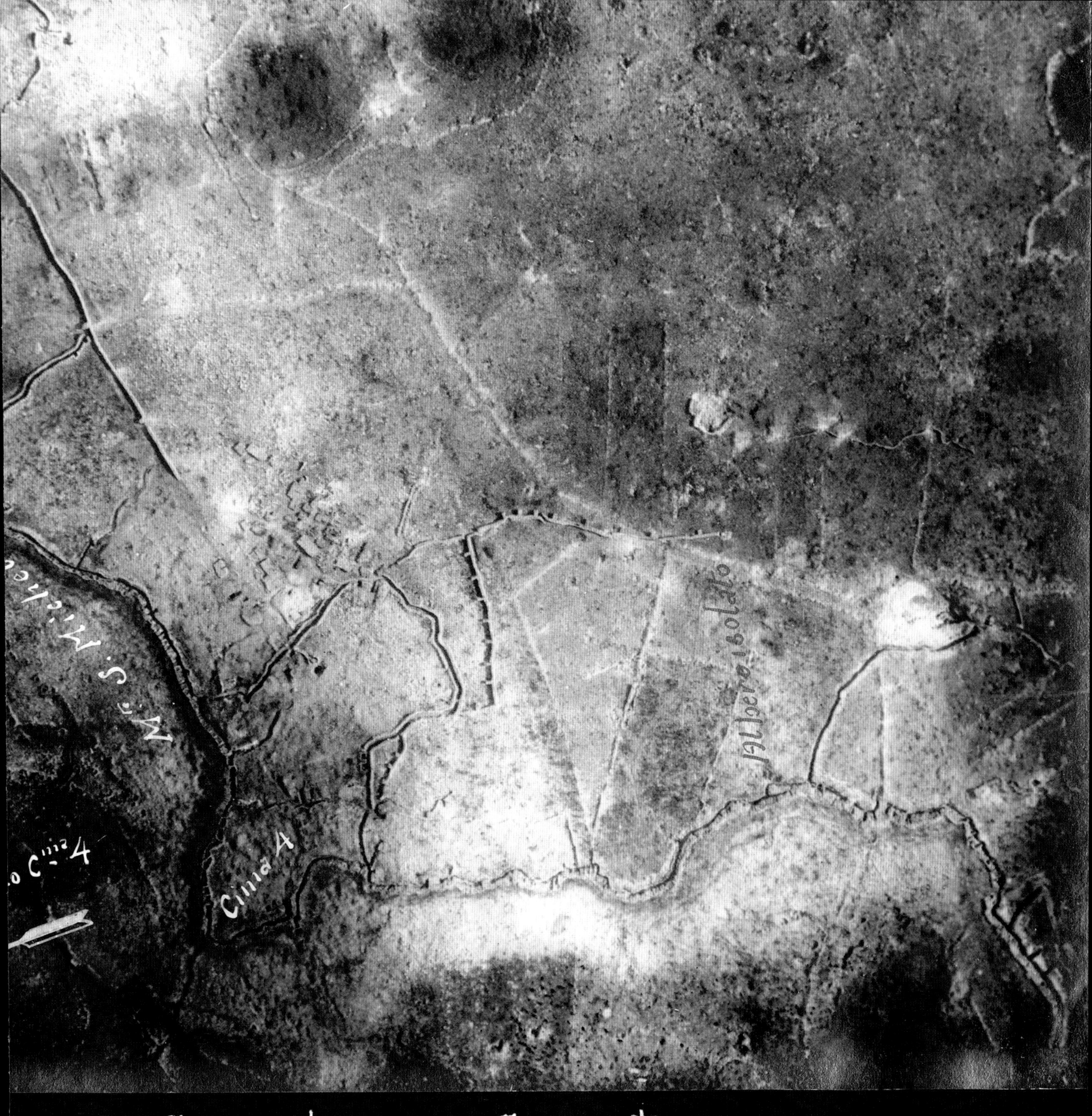

hes of Summit 3 and Summit 4 of Mt.

A Sopwith Camel—a British one-seater fighter plane of World War I—"comes a cropper" with its nose down in the French mud. British soldiers on horse and bicycle inspect the scene.

The Smithsonian took an early interest in preserving aircraft propulsion history. In 1889, Secretary Langley acquired a model-aircraft steam engine and two wood-and-fabric propellers designed by English experimenter John Stringfellow.

AERO PROPULSION

DOWTY R391 PROPELLER **1999**

HISPANO-SUIZA TIMING WHEEL **1918**

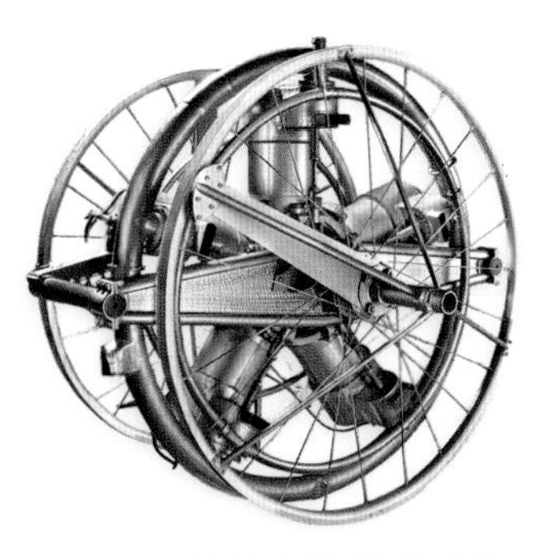

MANLY-BALZER RADIAL 5 **1903**

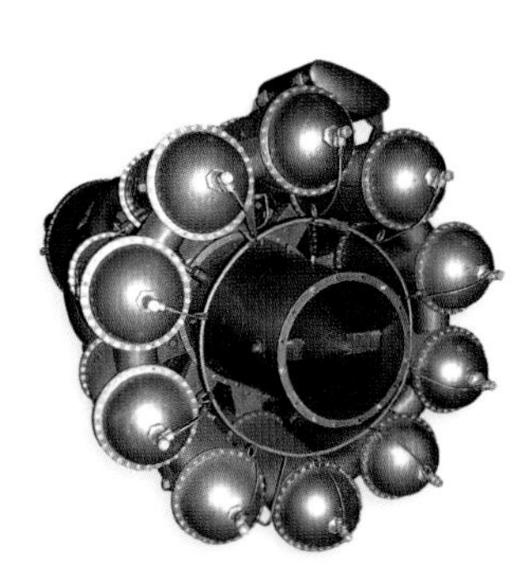

WHITTLE W.1X TURBOJET **1941**

PRATT & WHITNEY R-2800 **1943**

WRIGHT CYCLONE R-3350-23 **1944**

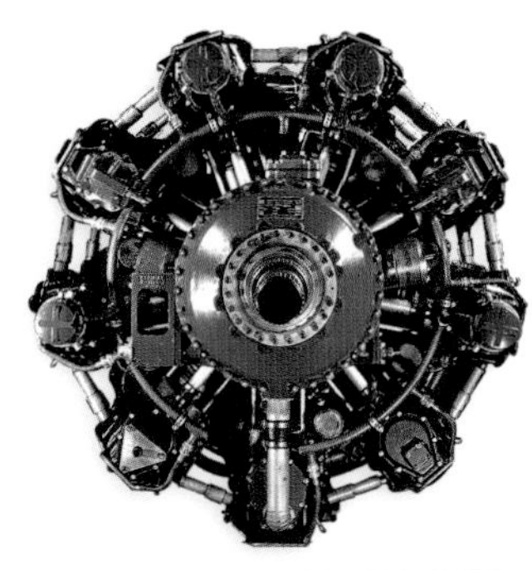

LYCOMING XR-7755-3 RADIAL **1945**

LE RHONE MODEL J **1917**

WRIGHT WHIRLWIND J-5 **1929**

LYCOMING T53 TURBOSHAFT **1960**

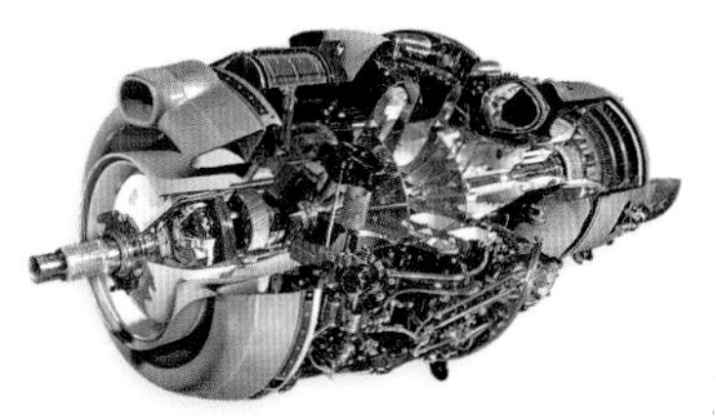

ROLLS-ROYCE DART MK. 520 **1970**

WRIGHT CYCLONE 18R-3350-TC **1952**

GENERAL ELECTRIC CF6 **1990**

PRATT & WHITNEY F100-PW-100 **1970**

TELEDYNE CAE J402-CA-400 **1972**

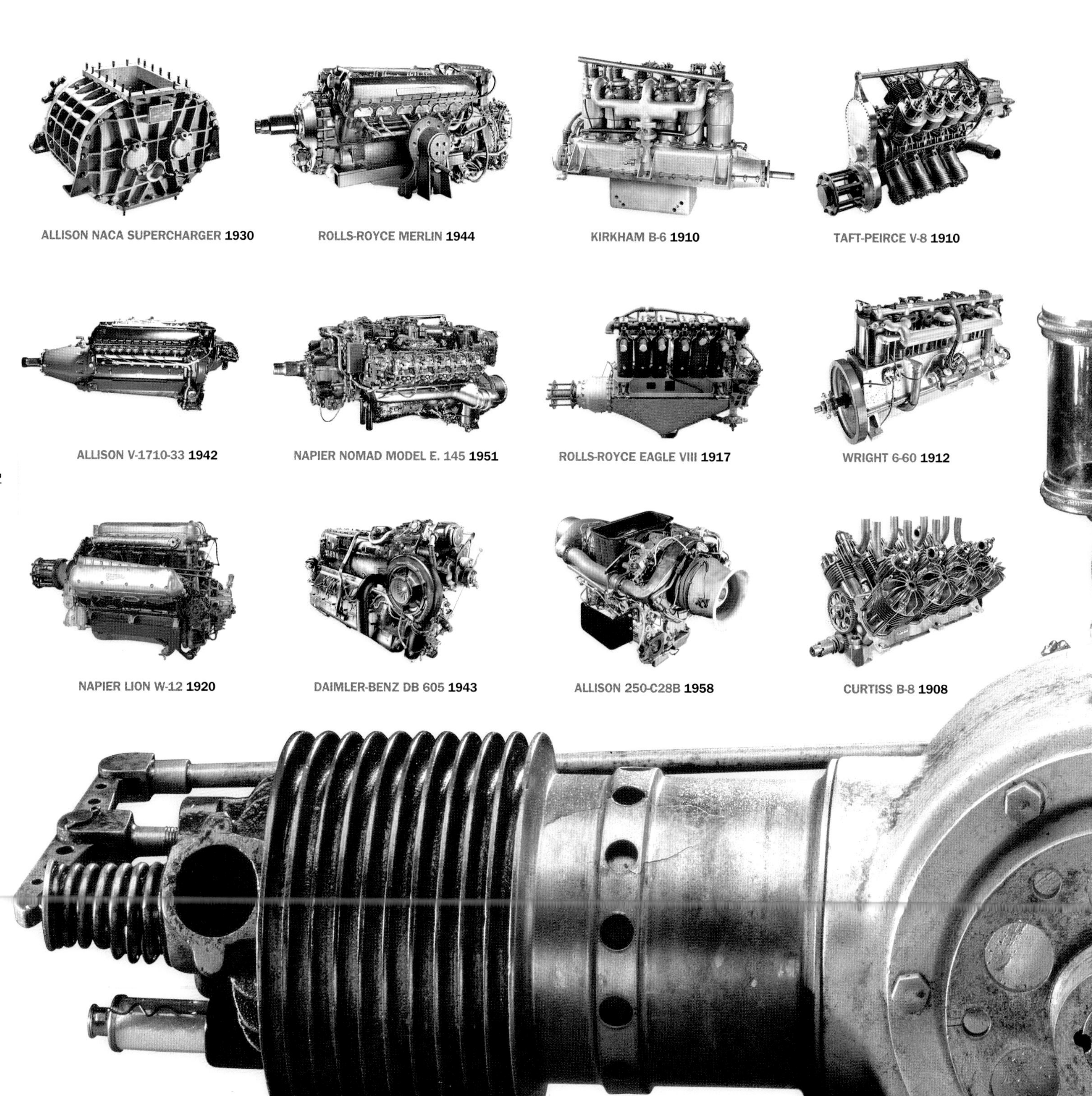

ALLISON NACA SUPERCHARGER **1930**

ROLLS-ROYCE MERLIN **1944**

KIRKHAM B-6 **1910**

TAFT-PEIRCE V-8 **1910**

ALLISON V-1710-33 **1942**

NAPIER NOMAD MODEL E. 145 **1951**

ROLLS-ROYCE EAGLE VIII **1917**

WRIGHT 6-60 **1912**

NAPIER LION W-12 **1920**

DAIMLER-BENZ DB 605 **1943**

ALLISON 250-C28B **1958**

CURTISS B-8 **1908**

Since then, the collection has evolved into the world's best, with over 1,700 individual artifacts dating from the very origins to the late 20th century and illustrating many approaches to improved engine performance. / For the first half of the century, the primary form was reciprocating or rotary engines joined to propellers, the basic system introduced by the Wrights in 1903. The invention of jet engines in the 1940s, pioneered by Sir Frank Whittle and Dr. Hans Pabst von Ohain, led to four new configurations—turbojet, turboprop, turboshaft, and turbofan—that again revolutionized aircraft use and performance. / The story of aeropropulsion is about a specialist community of inventors, engineers, workers, mechanics, and entrepreneurs interacting with the larger aeronautical community to maximize the power and performance of airplanes, helicopters, and lighter-than-air craft. These engines, propellers, accessory components, and miscellaneous parts represent their journey through flight.

LIBERTY 12 MODEL A **1917**

KING-BUGATTI U-16 **1918**

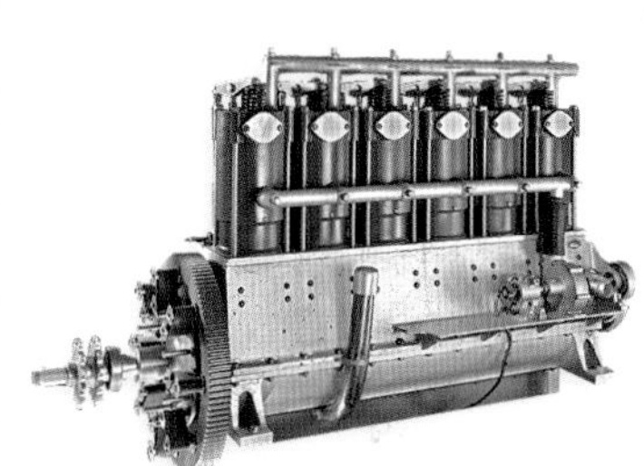
WRIGHT 6-70 **1913**

BATES MODEL 2 B **1910**

1.5 CAPABLE OF FLIGHT? THE WRIGHT-SMITHSONIAN CONTROVERSY

The Langley Aerodrome as rebuilt by Glenn Curtiss rests on the ice of Lake Keuka, Hammondsport, New York, in this 1914 photo.

harles Doolittle Walcott, the fourth Secretary of the Smithsonian, was determined to honor his friend and predecessor, Samuel Pierpont Langley. He installed a memorial tablet on the wall of the Smithsonian Castle, established the Langley Medal for contributions to aeronautics, created the Langley Aerodynamical Laboratory, and ensured that the NACA's first laboratory would be named in Langley's honor. May 6—the day on which the Smithsonian crew had flown the first successful steam-powered unmanned aerodrome—became Langley Day.

At Alexander Graham Bell's suggestion, Walcott presented the first Langley Medal to the Wrights, yet the brothers were always wary of the Secretary, with good reason. The problems began in 1910, when Walcott declined their donation of the 1903 Wright airplane to the Smithsonian. The Secretary requested "one of your machines, or a model thereof, for exhibition purposes." The Wrights offered to "reconstruct the 1903 machine with which the first flights were made at Kitty Hawk," and were stunned to hear that the Smithsonian would prefer the 1909 Military Flyer. Walcott also requested aircraft models and full-scale engines to display with specimens from the Langley collection, "making the exhibit illustrate two very important steps in the history of the aeronautical art." The Smithsonian planned to exhibit the 16-horsepower Wright engine of 1903 next to the 52-horsepower Langley engine; a 1909 Wright aircraft with parts of the 1903 Langley machine; and scale models of manned Wright aircraft with the unmanned Langley steam models of 1896.

But the real problems began in 1914, when Glenn Hammond Curtiss, an aircraft builder from Hammondsport, New York, whom the Wrights were suing for patent infringement, requested that the Smithsonian allow him to rebuild the sad remains of the 1903 Langley machine, the Great Aerodrome, and test-fly it to better understand the performance of tandem wing airplanes. Of course, the real reason for the request was the possibility of actually flying the old machine and showing that someone had been "capable of flight" before the Wrights. Anxious to demonstrate that Langley might have flown if only things had gone

Secretary Charles Walcott (left) and Alexander Graham Bell escort Orville (right) and Wilbur Wright (left center) to a waiting automobile following their award of the Langley Medal on February 10, 1910. OPPOSITE: Glenn Hammond Curtiss (left) and Albert Francis Zahm consult during the 1914 tests of the rebuilt Langley Aerodrome.

The Langley Aerodrome as rebuilt by Glenn Curtiss rises from the water of Lake Keuka during the 1914 tests. OPPOSITE, TOP: Glenn Curtiss sits at the control of the *June Bug*, the aircraft in which he won the Scientific American Trophy for a straight-line flight of one kilometer on July 4, 1908. OPPOSITE, BOTTOM: Glenn Curtiss (left) and Albert Francis Zahm pose with the Langley Aerodrome, radically altered for the 1914 flight tests.

a bit differently, Walcott jumped at the chance. Without informing the Board of Regents, he authorized Albert F. Zahm, head of the Langley Aerodynamical Laboratory, to turn over all the surviving parts to Curtiss.

If the goal was to return the aerodrome to its 1903 condition, Curtiss and Zahm failed. The wings constructed in the Curtiss plant differed from the originals in chord (the straight-line distance from the leading edge to the trailing edge), camber (distance from the peak of the arch of the wing to the imaginary chord line), and aspect ratio (the ratio of span to chord). The trussing system that linked the wings to the fuselage was also quite different. This change was particularly important, for most authorities believed that the failure of the wing structure, not a catapult defect, had been responsible for the 1903 disaster.

There were other changes as well. Curtiss fitted the craft with his own yoke-and-wheel flight control system. He moved the pilot to a new position on top. Thus, the pilot would not end a successful flight underwater, as would have been the case in 1903. After the first trial, he altered the tail to serve as both rudder and elevator. Finally, he rejected the old catapult launch system, mounting the machine on floats. This change can be excused in the name of simple self-preservation. It does not seem to have occurred to anyone at the time, however, that Curtiss had come up with a way to land the machine safely, which had been impossible with the original craft. In short, he had transformed a machine that had failed to fly in 1903 into a less-than-satisfactory 1914 airplane.

On the morning on May 28, 1914, the rebuilt aerodrome, with Curtiss at the controls, sped across the surface of Lake Keuka, near the site of the Curtiss factory in Hammondsport, New York, and lifted into the air for a flight of 150 feet. After a few additional hops of similar length, the craft was taken back into the shop, where the 1903 Langley engine was replaced with a modern Curtiss power plant, and additional flights were made.

Walcott and Zahm were overjoyed. In an account of the tests published in 1914, Zahm claimed that the aerodrome "has demonstrated that with its original structure and power, it is capable of flying with a pilot and several hundred pounds of useful load. It is the first aeroplane in the history of the world of which this can truthfully be said." He lied that it had flown "without modifications." "With a thrust of 450 pounds," he concluded, "the Langley aeroplane, without floats, restored to its original condition and provided with stronger bearings, should be able to carry a man and sufficient supplies for a voyage lasting practically the whole day." The 1915 annual report repeated the claim. When the aerodrome was shipped back from Hammondsport, Walcott ordered that it be returned to its original 1903 condition and displayed in the Arts and Industries Building, labeled as the "first man-carrying aeroplane in the history of the world capable of sustained free flight."

Orville Wright was justifiably outraged. Plus, there was the obvious conflict of interest, because the Smithsonian had sponsored the reconstruction and testing of the Langley machine by a man with whom the Wrights were locked in a bitter patent fight. In 1921 Griffith Brewer, an English friend of Orville's, cataloged the changes made to the aerodrome during the first and second episodes of rebuilding at Hammondsport. The evidence was overwhelming: The 1914 tests had not demonstrated that the 1903 Langley Aerodrome was "capable of flight."

Orville Wright remained aloof from the controversy until 1921, unwilling to make what might be seen as "a jealous attack upon the work of a man who was dead." When his private letters of protest to Smithsonian officials, including Chief Justice William Howard Taft, were ignored, however, he decided to take direct action. On April 30, 1925, Orville Wright played his trump card, announcing that he was sending the 1903 Wright airplane to the Science Museum of London, where it would remain in exile unless and until the Smithsonian corrected its false statements. The announcement galvanized public attention. Attempts to persuade the inventor of the airplane to change his mind were to no avail. The crates containing the restored 1903 Wright aircraft were loaded aboard the *Minnewaska,* which sailed for England on February 11, 1928.

Orville Wright's decision to send the world's first airplane to an English museum put Secretary Walcott on the defensive, but he refused to budge. Following Walcott's death in 1927, his successor, Charles Greeley Abbot, reduced the label on the aerodrome to read: "Langley Aerodrome—The Original Langley Flying Machine of 1903. Restored." The following year the Smithsonian Board of Regents passed a resolution declaring, "To the Wrights belongs the credit of making the first successful flight with a power-propelled heavier-than-air machine carrying a man." Orville Wright was unmoved, insisting that the only resolution to the problem would be a published statement by the Smithsonian detailing differences between the original aerodrome and the aircraft flown at Hammondsport in 1914, together with an honest admission that the tests had not demonstrated the capacity of the Langley Aerodrome for flight.

The Smithsonian feud with the Wright brothers was now the longest running controversy in the short history of aviation. The tide was turning against the Institution. Finally, on October 24, 1942, following another series of discussions, Abbot published an article giving Orville Wright exactly what he had asked for. Orville Wright did not respond, however. He died on January 30, 1948, without having informed the Smithsonian of the ultimate fate of the 1903 Wright airplane.

As the executors of the Wright estate were to discover, however, Orville had decided that the publication of Abbot's 1942 article *was* satisfactory. He had written on December 8, 1943, to inform the director of the Science Museum that he would be asking for the return of the machine once the war was over and the craft could safely be transported. This letter and the provisions of an unsigned will made clear that Orville Wright intended the world's first airplane to reside at the Smithsonian.

The airplane's 20-year exile ended on the morning of December 17, 1948, when the 1903 Wright Flyer was unveiled in the North Hall of the Smithsonian's Arts and Industries Building. Sir Oliver Franks, the British ambassador to the United States, summed the moment up when he looked up at the machine and commented that it was "a little like being in the presence of the original wheel."

Was the 1903 aerodrome ever capable of flight? A study by NASA engineers in 1982 pointed to the torsional weakness of the 1903 wings, and in 1998 Lorenzo Paul Joseph Auriti concluded: "The Langley Aerodrome was not structurally capable of level flight since the wings could not support aerodynamic loads." — *Tom D. Crouch*

This photograph, perhaps the most famous in aviation history, was taken at about 10:35 on the morning of December 17, 1903. It shows the world's first airplane taking off for the first time. Orville Wright is piloting the machine, while Wilbur runs along beside.

2 BUILDING A COLLECTION

F. ROBERT VAN DER LINDEN

Spirit of St. Louis

MILESTONES OF FLIGHT

With the end of the Great War, *the*

toward building a significant aviation collection. As one of the victors in World War I, the U.S. government was eager to display its military prowess to the American people. While professional museums were a relatively new idea that grew enormously during the late 19th and early 20th centuries, the idea of a technological museum was even newer. Yet people clamored for something more permanent and interpretive than the promotional displays of machines of the industrial age in countless expositions. / One of America's most prominent aviation advocates, Edward P. Warner, wrote in support of the creation of a national aeronautical museum to educate professionals as well as the lay public.

Smithsonian began to take its first steps

Britain had its Science Museum in London and France had its Conservatoire National des Arts et Métiers in Paris, but few other countries outside Europe displayed their scientific and engineering artifacts. The Smithsonian Institution in Washington was the only museum of this type in the United States, but it could display only a small amount of the nation's mechanical history. A national museum dedicated to aeronautics should be more than a limited collection of "highpoints in the history of the art," Warner wrote. He envisioned a museum where actual aircraft could be displayed for study by engineers and where different technological applications could be examined and compared for future use.

Loening OA-1A ***San Francisco***
1926

FROM THE MUSEUM'S COLLECTION

1920s TO 1940s

Douglas World Cruiser ***Chicago***
1924

Boeing P-26A Peashooter
1934

Bellanca C.F.
1922

Lockheed Vega 5B
1932

Ryan NYP *Spirit of St. Louis*
1927

Lockheed Sirius *Tingmissartoq*
1929

Bell XP-59A Airacomet
1944

V-2
Ca 1945

Dornier Do 335 A-0 Pfeil (Arrow)
1945

Messerschmitt Bf 109 G-6/R3
1944

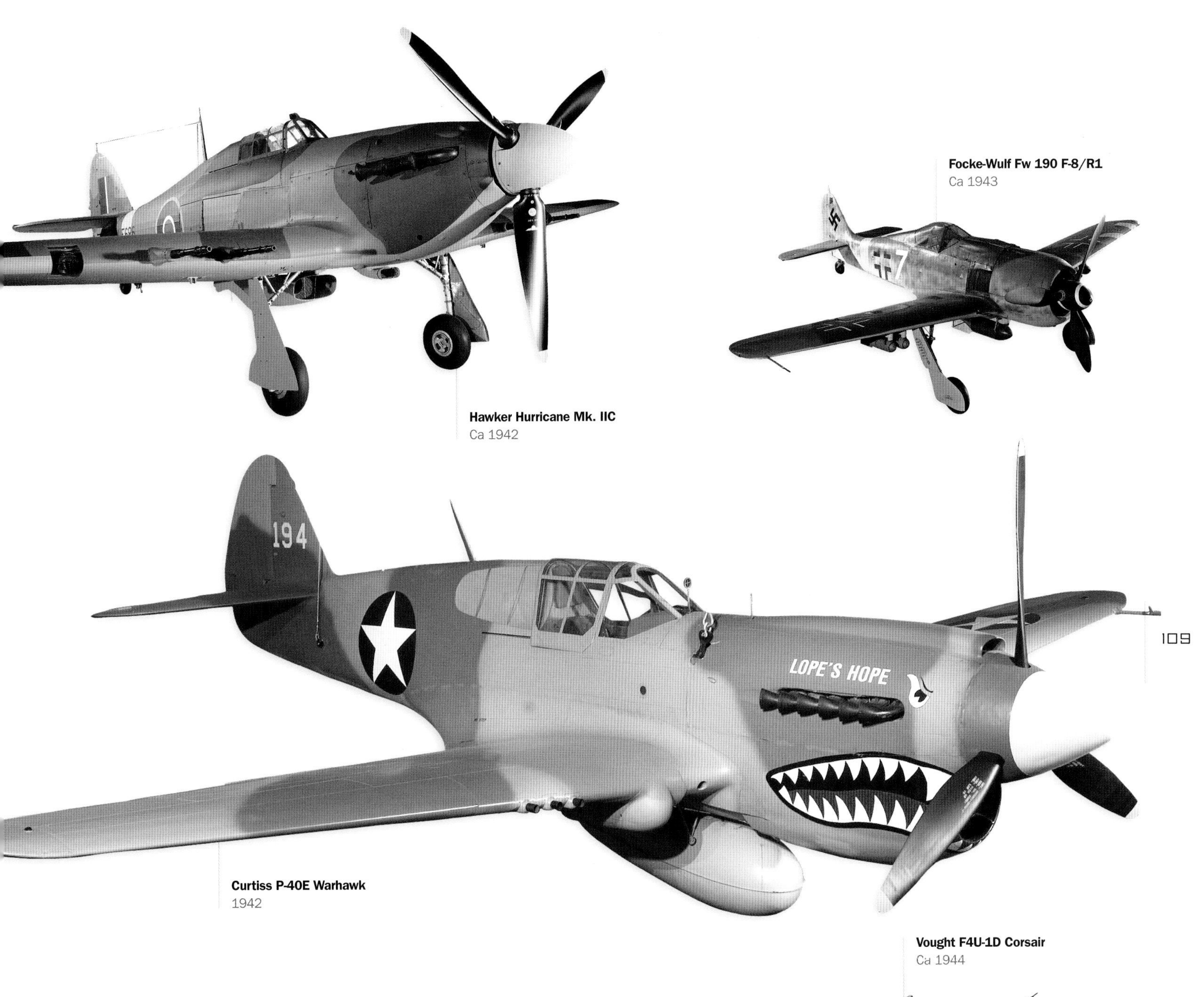

Focke-Wulf Fw 190 F-8/R1
Ca 1943

Hawker Hurricane Mk. IIC
Ca 1942

Curtiss P-40E Warhawk
1942

Vought F4U-1D Corsair
Ca 1944

Curtiss F9C-2 Sparrowhawk
1932

Such a museum should display "many models which illustrate the continuous development of airplane design, although the machine which they represent may not have to their credit any single epoch-making feat," explained Warner. A collection of well-labeled models "is very useful instructing the non-aeronautical public and is likely to prove of greater service in the long run than the holding of an annual air show lasting only a few days and presenting only new types." The Smithsonian Institution would be a logical place for such a museum.

Faced with the arrival of a significant, though temporary, collection of World War I military aircraft from the U.S. Army and Navy, the Smithsonian scrambled to find a home for them. With few resources and no staff members who understood this new technology or were willing to dedicate themselves to its collection and preservation, the Institution did what it could. The idea of a national museum dedicated to aeronautics would have to wait.

The first solution was literally in its own backyard. During the war, between June and October of 1917, the government erected a temporary $28,000 metal building on the grounds of the Smithsonian, behind the Castle and close to Secretary Langley's old workshop. The building was first used to test Liberty engines for a series of U.S.–built military and naval aircraft. After June 1918, the building served as a warehouse for stationery supplies for the War Department. In 1919, the building was transferred to the Smithsonian and rebuilt with siding, insulation, improved lighting, new

Built in 1918, the Aircraft Building housed most of the museum's aviation collection for decades. Taken in 1920, this photo also shows a tank and artillery piece displayed by the front door.

PAGES 102–103: Paul E. Garber poses in front of a major acquisition, the Ryan NYP *Spirit of St. Louis,* soon after the new National Air and Space Museum opened in 1976.

PARK HEAD IN
TWO HOURS
NO PARKING
ENTRANCE
428-59

paint throughout, and a concrete floor, which replaced the original wooden planking. It was quickly filled to capacity with new aircraft and other World War I artifacts. For a time the renamed Aircraft Building housed a huge Martin bomber, a LePere fighter-bomber, an Aeromarine 39B floatplane, and a skeletonized de Havilland DH-4 light bomber, all transferred from the War and Navy Departments.

The second solution walked through the door the next year. A slight, cheerful young man, Paul Edward Garber visited the National Museum's Arts and Industries Building in early June 1920. Noticing that the control wires of the historic 1909 Wright Military Flyer had been incorrectly installed, he approached the responsible curator about it. But instead of a condescending rebuke, the curator offered the upstart 20-year-old a job as an exhibit preparator. Garber used his three-month appointment wisely and was offered a permanent job at the end of the summer, once he turned 21. Garber seized his opportunity and took on the responsibilities of handling this new collection.

OPPOSITE, TOP: Paul Garber shows his rubber-powered model aircraft to Cmdr. John Rodgers and his family at Bolling Field in 1924.

OPPOSITE, CENTER: Garber demonstrates Otto Lilienthal's original hang glider for the 1936 British film *Conquest of the Air.*

OPPOSITE, BOTTOM: Model maker Edward William Topping (left) and Garber examine two models that Topping donated in 1930.

LEATHER COAT 1920s Worn by Amelia Earhart on many of her flights

PAUL EDWARD GARBER

Who was Paul Garber? From 1920 until his death in September 1992, Paul Garber was the heart and soul of the Smithsonian aviation collections and, later, the National Air and Space Museum. A personable dynamo who stood just over five feet tall, Garber was a giant in the world of aviation museums, collecting most of the Smithsonian's outstanding aircraft. Unfazed by the limited support he received, Paul Garber used his wits and imagination to persuade, plead, and cajole donors to part with their artifacts—doing whatever it took to enhance the national collection. Through his tireless efforts, he collected and preserved America's priceless aeronautical treasures for future generations to enjoy—until his efforts were finally fulfilled with the creation of the National Air Museum, as it was originally called.

With a ready smile and an amusing quip, Garber not only relentlessly pursued aircraft for the collection, he also used his myriad connections to help establish the legal framework for the creation of the museum. Despite decades of lukewarm support from his superiors, Garber remained faithful to the Institution and to James Smithson's original bequest to build "an Establishment for the increase and diffusion of knowledge among men."

Paul Edward Garber was the son of Paul Greenwood Garber, an art dealer and interior decorator, and Margaret Sithins Garber. He was born on August 31, 1899, in Atlantic City. At the age of five, he was introduced to the wonders of flight by his uncle Edward, who taught him to fly a kite along the Atlantic shore. Ed had secured the kite to young Paul's wrist to keep it from blowing away. It almost achieved the opposite purpose. "I trotted out into the waves, headed for Europe," he was later quoted as saying, "totally thrilled by the lift of that kite."

In the summer of 1909, Paul Garber's father took his ten-year-old son with him when he traveled to Washington, D.C., to see the sights and explore business opportunities. While perusing an edition of the evening *Star,* Paul Garber happened upon an announcement that Orville Wright was to demonstrate his Military Flyer the next day at Fort Myer, Virginia, across the Potomac. Financed with 50 cents from his father, Paul hopped the trolley to Arlington, about a mile from the fort. He ran the rest of the way.

He knew that he was in the right place. As Paul approached the gate, he could hear the strange sound of that Wright Flyer warming up at the end of the parade ground. Seventy-two years later, in an interview with *Johns Hopkins Magazine,* the memory was still fresh. He was transfixed, he said, "to see an enormous thing like that, much, much greater than any kite, with two men in it, and no kite strings attached, and flying towards me from the far end of the field, with propellers churning." "I saw it go up in the air," he continued, "and followed it until I fell over backwards and I just lay there in amazement."

Not long after his eye-opening experience at Fort Myer, Garber and his family moved to Washington. Paul's father had found a suitable location for his gallery on the ground floor of a Connecticut Avenue building. Upstairs on the second and third floors, the Garbers rented a large apartment. Today a crowded urban area, this part of Washington still had open fields in 1910, a perfect location for a young boy to enjoy flying his kites.

Up the street from the Garbers lived one of the world's great inventors and aeronautical enthusiasts, Alexander

Graham Bell. Though best known for his pioneering work with the telephone, Bell was an accomplished scientist who actively conducted aeronautical research with Glenn Curtiss and the Aerial Experiment Association. Like his ten-year-old neighbor, Bell was particularly interested in kites. As the story goes, one day Garber was playing on Connecticut Avenue when Bell and a friend walked by and noticed his simple kite. To Paul Garber's astonishment, Bell walked over, announced that the bridle was incorrectly assembled, and hauled down the offending kite. Swiftly Bell handed the boy the kite, removed his pocketknife, and cut off the bridle. He then cut a length of string, tied a new bridle, stepped into the street, and with a strong breeze at his back, skillfully lifted the kite back into the air. Bell stood beside Garber admiring his handiwork for a moment before returning the kite and patting him kindly on the head.

Flight in all its forms now had the young Garber in a firm grip. When he had the time, he would ride his bicycle out to a College Park airfield in suburban Maryland to watch America's first military pilots practice their trade. This included the Army's 13th pilot, First Lt. Henry H. "Hap" Arnold, future leader of the Army Air Forces in World War II and the first and only five-star general of the Air Force. Arnold would later play an important part in Garber's professional life.

In 1913, Paul Garber organized the Capital Model Aero Club, which sponsored kite and flying-model competitions. Two years later, while visiting the Smithsonian's National Museum, he happened upon a scale model of aerial pioneer Octave Chanute's biplane glider. He was | to page 116 |

Famed inventor and aviation pioneer Alexander Graham Bell poses with a child sitting on a pontoon of a Sikorsky S-38 during the 1930s.

| from page 113 | inspired to rush home to build his own kite version, which flew quite nicely. The thought then occurred to Garber: If it flew so well as a kite, it might do as well in its original form. The hope of an impressive view from a perch in the sky was enough to convince him that he should build his own glider.

Paul Garber assembled the necessary materials: sawed barrel staves provided the structure, red chintz from his mother provided the fabric, and great lengths of clothesline from the neighbors braced the craft and acted as a tow line. Several times, he managed to lift off the ground.

In this venture, he was helped by his diligent friends from the model club, who charged along in an open field near California Street and Massachusetts Avenue, pulling Garber with the kite firmly on his shoulders. He estimated that he flew about a dozen times with this help and in free flight. His best flight occurred in August 1915, when he estimated he flew about 400 feet from a hill at the end of the field.

For reasons that he never explained, Paul Garber did not persevere with gliding much longer after these early efforts, although he did help to organize the Washington Glider Club and assisted Catholic University aeronautics professor Louis Crook in building two gliders. World War I was approaching, and he soon turned to other pursuits.

Garber was too young to enlist when the United States entered the First World War in 1917 but took part in the war effort as best he could. His parents thwarted his initial attempt to enlist in the Navy; however, he was permitted

TOP: Mail is loaded into a Douglas M-2 before the opening flight of Western Air Express along Contract Air Mail Route 4 (CAM 4) between Salt Lake City and Los Angeles, April 17, 1926.

BOTTOM: Ford 5-AT Tri-Motors could store mail, packages, and baggage inside thick cantilevered wings. Here a Ford of Transcontinental & Western Air (TWA) is loaded in Kansas City in the early 1930s.

OPPOSITE: U.S. Air Mail pilot William C. "Wild Bill" Hopson strikes a cocky pose before a flight from Omaha to Chicago in 1921. Hopson earned a reputation for disobeying Post Office flight regulations. He died in a crash while flying the mail for National Air Transport on October 18, 1928.

PILOT Wm. C. HOPSON
U.S. MAIL SERVICE
WINTER FLYING CLOTHING

to enlist in the Boys' Working Reserve. In this program, which was run by the Department of Agriculture, youths worked on a farm in Kimberton, Pennsylvania, during breaks from school. After his 18th birthday in August 1917, Garber enlisted in the District of Columbia National Guard, and he eventually transferred to the U.S. Army.

With the Army's assistance, he took ground school at Maryland State College and received several flight lessons at Bolling Field from Maj. George Burgess in the ubiquitous Curtiss JN-4 trainer. Garber had known Burgess previously through the Capital Model Aero Club. The armistice on November 11, 1918, brought an end to his flying lessons, and one month later he received an honorable discharge from the Army, having attained the rank of sergeant. Ironically, though, for the future head curator of the National Air and Space Museum, he never completed his ad hoc flight training, nor did he receive a pilot's license.

Garber was stationed in the Washington area during the war and witnessed America's first practical steps toward commercial aviation. When the U.S. Post Office Department initiated regular air mail service between Washington and New York on May 15, 1918, Paul was among the well-wishers at the Polo Grounds near the Potomac. He watched Lt. George Boyle take off for points north in a Curtiss JN-4 and stayed until Lt. James Edgerton flew in from Philadelphia with the southbound mail. Three days after his discharge from the Army, and 15 years to the day after the Wright brothers flew the first powered heavier-than-air craft, Paul joined the U.S. Air Mail Service.

OPPOSITE: Lt. James H. "Jimmy" Doolittle stands in front of his de Havilland DH-4 of the 90th Squadron near San Antonio, Texas, in September 1922.

FLIGHT WINGS Ca 1938 Diamond-encrusted gold wings with the initials "RT" for Roscoe Turner in red enamel

Once the Air Mail Service became well established, the Post Office moved its operation a few miles north to the Army's larger field at College Park, Maryland, which survives today as the world's oldest continuously operating airport. Garber served there as a jack-of-all-trades, working as a mechanic, a tool and stock clerk, and a driver who picked up air mail at branch offices throughout Washington. Occasionally he would fly as a passenger during test hops and would receive impromptu lessons when time permitted. On Independence Day 1919 he even soloed a Curtiss Jenny under the watchful eye of Major Burgess.

Garber's career with the Post Office was productive but brief. The authorities soon realized that the air mail route between Washington and New York was too short to realize any speed advantage over the existing train service. As the goal was to link the East and West Coast of the country, the Post Office Department opened a new route between New York and Chicago as the first step in that plan. This was achieved by the late spring of 1919. Garber was to be transferred to Chicago, but the sudden illness of his father forced him to resign so that he could run the family business until the senior Garber recovered.

Paul Garber soon realized that the world of art was not for him; he simply loved airplanes too much. In the postwar atmosphere, there was little opportunity for civilian aviation careers, but he had to do something. During this time he would often visit the Smithsonian, especially to see the Wright Military Flyer, which was the same aircraft that inspired him a decade before.

It was during one of these visits that he was offered the job as preparator, responsible for repairing and maintaining exhibits. His summer job began on June 1, 1920, with a host of small projects. On his own, he undertook to prepare a mannequin for the Military Flyer to help visitors understand how the machine flew. Using his own Army uniform, Paul prepared and installed the "pilot" after the curator had left for vacation in August.

With his appointment quickly running out, Garber hastened to finish his other project, that of building a model of Leonardo da Vinci's ornithopter (flapping-wing machine). He was atop a ladder busily installing the model with an explanatory label when an inquisitive elderly man approached him. He spent the next 30 minutes regaling the gentleman about da Vinci and his myriad scientific contributions. When asked who he was, Garber replied that he was just a temporary worker whose appointment had already expired, but he wanted to finish his model before leaving. With that, the elderly man departed, and he went about cleaning up.

As Garber subsequently told the *Washington Post*, "I put my ladder away and looked back at my model and went up to the shop and felt sorry for myself. . . . Up came the chief clerk of the Institution. Said, 'Your name's Garber? You built this thing without permission I understand?' I said, 'Yes, but I'm leaving | *to page 124* |

COPYRIGHT BY
H.L. SUMMERVILLE
SEPT. 6-1922

Today the Fokker T-2 is displayed in the museum's Barron Hilton Pioneers of Flight gallery.

The Fokker T-2 was the first airplane to make a nonstop flight across North America. U.S. Army Air Service Lt. Oakley G. Kelly and Lt. John A. Macready took off from Long Island, New York, on May 2, 1923, and landed at San Diego, California, slightly more than 26 hours and 50 minutes later.

FOKKER T-2

Designed and built in the Netherlands as the F.IV ten-seat transport, two of these aircraft were purchased by the Air Service in June 1922 and redesignated as T-2s.

The largest Fokker aircraft up to that time, the T-2 featured a fully cantilevered wooden monoplane wing spanning nearly 82 feet, a scaled-up version of the one on the Fokker D.VIII fighter used by the Germans late in World War I. An American-built 420-horsepower Liberty V-12 engine powered the T-2.

Modifications for the flight included two additional fuel tanks, which expanded the airplane's gasoline capacity from 130 gallons to 725 gallons. The Fokker normally had a single pilot's position in an open cockpit left of the engine. The T-2 was modified with a second set of controls in the cabin to facilitate the crew switching positions.

The first two coast-to-coast attempts started from San Diego, to take advantage both of the prevailing westerly winds and the refined fuel available in California, which had a higher octane rating. Both of these attempts failed.

A successful attempt was made in the reverse direction. Lieutenant Kelly was at the controls when they took off from Roosevelt-Hazelhurst Field in New York, on Long Island, at 12:30 p.m. on May 2. The crew switched positions several times. Macready landed the T-2 in San Diego on May 3 at 12:26 p.m. They had flown 2,470 miles at an average ground speed of 92 miles per hour.

In January 1924, the Air Service transferred the Fokker T-2 to the Smithsonian Institution. It was fully restored during the years 1962–64, and it received further minor repairs in the years 1973 and 2009.

The Fokker T-2 flies over the American countryside on its way to San Diego. **OPPOSITE, TOP:** After their historic 1923 transcontinental flight, Lt. John A. Macready (left) and Lt. Oakley G. Kelly (right) pose with Orville Wright. **OPPOSITE, BOTTOM:** The Fokker T-2 is pictured in flight over farmland.

| from page 118 | right now, sir.' He said, 'Can you take a civil service examination?' I thought I was going to get fired or put in jail. He said, 'Secretary's orders.' I said, 'The Secretary? Charles Walcott? He doesn't know me at all, I've never seen him.'"

Garber was mistaken. The elderly gentleman to whom he had spoken about da Vinci was, as a matter of fact, the Secretary. His temporary position now became quite permanent.

For the next 50 years Paul Garber remained in the Smithsonian's employ. During that time, which coincided with the greatest developments in aviation, he led the way in creating the finest collection of aeronautical artifacts in the world. He began almost immediately, even though he was virtually alone in his quest.

ASSEMBLING A WORLD-CLASS COLLECTION

As preparator, Garber worked for the princely sum of $700 a year. He was assigned to the Division of Mineral and Mechanical Technology, where he worked under curator Carl Mitman and a small staff. He spent most of his time building models, drawing plans, and constructing exhibits. On his own initiative, and usually in his spare time, he took a special interest in the museum's small, but important, collection of aircraft.

Throughout the 1920s, Garber assembled, virtually single-handedly, an impressive collection, particularly of military aircraft. In 1924, he had collected the Fokker T-2, the first aircraft to fly across the United States without stopping. Despite its massive 80-foot wingspan, Garber managed to squeeze the T-2 into the Aircraft Building. He collected documentation on the U.S. Navy's latest dirigible, the U.S.S. *Shenandoah.* He also brought in the experimental Berliner helicopter of Emil Berliner, a local experimenter who had flown successfully from College Park, Maryland.

By late 1924, the collection was increased with the acquisition of the unique Martin K-III Kitten, a tiny biplane intended for military use that incorporated a retractable landing gear, the first of its kind. And when the aircraft were not available, he filled the exhibition halls with detailed models, often of his own construction, usually from the Capital Model Aero Club.

In December 1924, Paul Garber's diligence and dedication were rewarded with his promotion from aide to assistant curator. His pay jumped to $2,400. (An earlier attempt by Carl Mitman to establish a separate section for aeronautics under Garber failed to receive Smithsonian support.) As his status and reputation grew, he began to travel to aeronautical events on behalf of the Smithsonian. In October 1925, he attended the Pulitzer Trophy Race at Mitchel Field, Long Island, where he saw Lt. Cyrus Bettis take the cup while piloting his Curtiss R3C-1 landplane to victory.

Two weeks later, Garber traveled to Baltimore, where the United States was hosting the prestigious international competition for the Schneider Cup. Flying the same aircraft, only now equipped with floats and designated the R3C-2, Army Lt. Jimmy Doolittle triumphed against | to page 128 |

OPPOSITE: The museum houses thousands of drawings and blueprints of aircraft, including this one, representing the Curtiss CR-2 Racer.

STUFFED TOY MONKEY 1924 "Maggie" the monkey, mascot of the Douglas World Cruisers

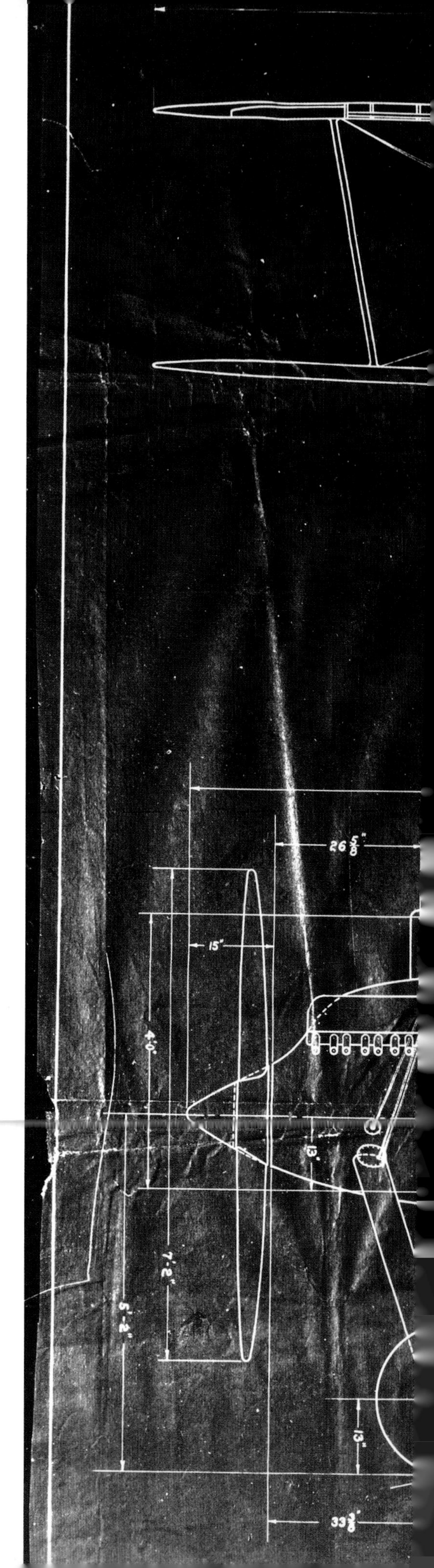

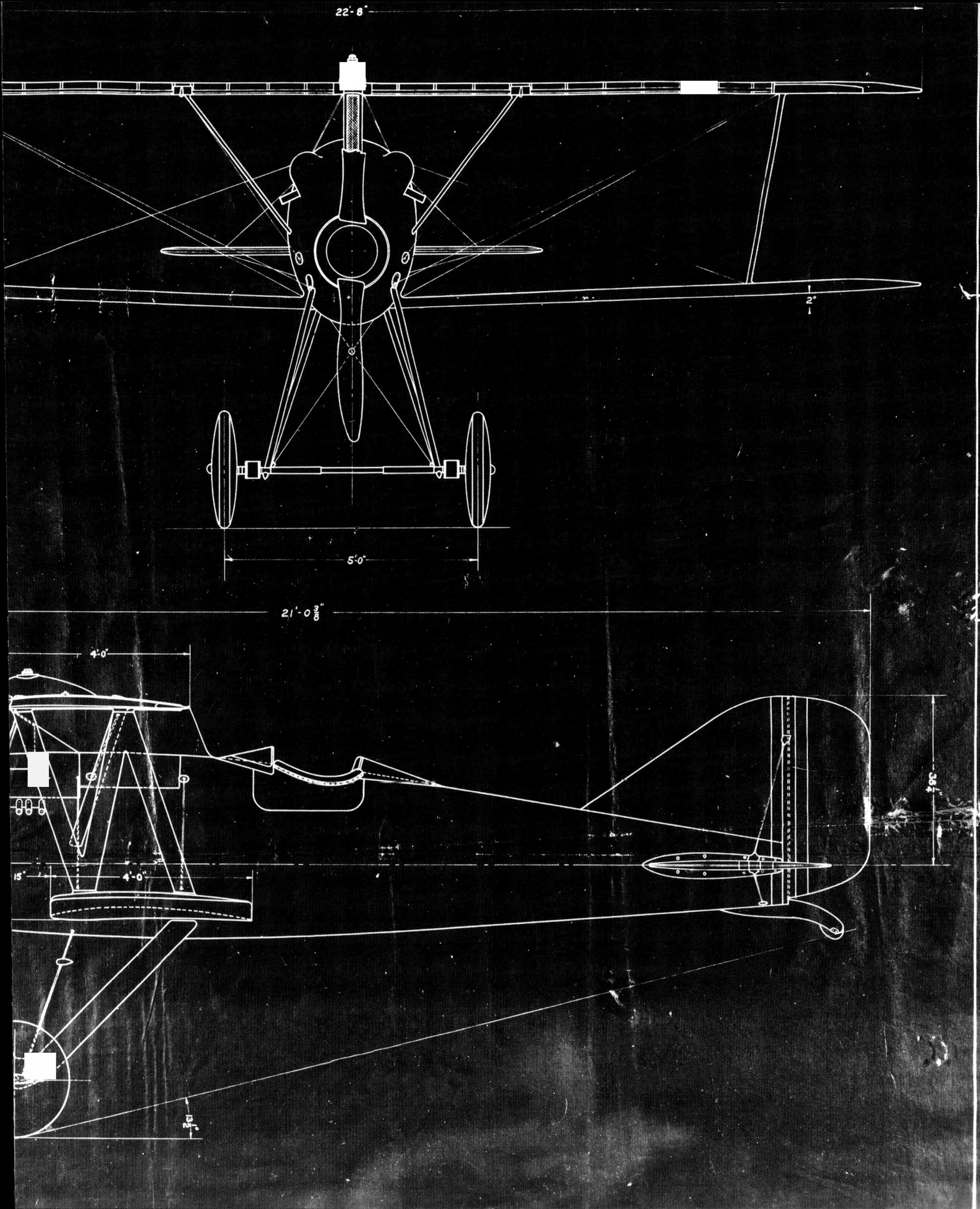

22'-8"
2°
5'-0"
21'-0 3/8"
4'-0"
15"
4'-0"
13½"

All four Douglas World Cruisers flew over Resurrection Bay, Seward, Alaska, before continuing on their voyage. BELOW, LEFT: After the crash of the World Cruiser *Seattle* in Alaska, the remaining three aircraft flew on, landing in Kagoshima, Japan, on June 2, 1924. BELOW, LEFT CENTER: The three Douglas World Cruisers refueling in Houton Bay, Orkney Islands, Scotland, on August 2, 1924. BELOW, RIGHT CENTER: On September 8, 1924, the World Cruisers make their triumphal entry into New York City. BELOW, RIGHT: President Coolidge congratulates Capt. Lowell Smith when the World Cruisers reach Washington, D.C., on September 9.

World Flyers arriving at Kagoshima, Japan, June 2.

1561 A.S.

World Flyers arriving at New York City, September 8

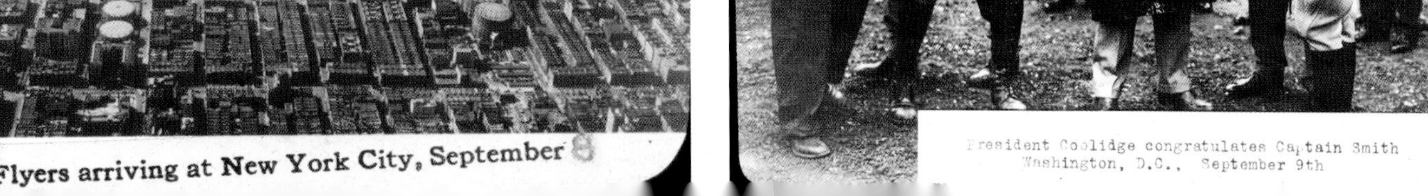
President Coolidge congratulates Captain Smith
Washington, D.C., September 9th

| from page 124 | the best aircraft Europe had to offer. Inspired by what he saw, Garber eventually arranged with the Army for this famous aircraft to join the Smithsonian in late 1927.

Soon after he returned from Baltimore, the War Department transferred the Douglas World Cruiser *Chicago* to the museum. This important acquisition of the first aircraft to circumnavigate the Earth, along with three other similar aircraft in 1924 (two of which survived), led to another rearrangement in the Aircraft Building. By far the most significant aircraft they had acquired up to that time, the *Chicago* replaced the Packard LePere light bomber that had been on display since the end of the war. The LePere, along with the Martin bomber and the de Havilland DH-9 and other U.S. Army aircraft, were all eventually returned to the military when new aircraft were acquired. Garber had always understood that these aircraft were to be held in storage until the time when the Smithsonian could expand and retrieve them. Unfortunately, the Army failed to live up to this agreement, and they were later declared surplus and destroyed, much to his eternal chagrin.

Through his acquaintance with Bill Moffett, the son of Rear Adm. William A. Moffett, the first chief of the Bureau of Aeronautics and the father of naval aviation, Paul Garber was able to persuade the Navy to preserve the Curtiss NC-4, the first aircraft to fly across the Atlantic Ocean. The Aircraft Building was too small to house the massive NC-4 flying boat. Following the 1926 Sesquicentennial Exposition in Philadelphia, where the NC-4 was displayed, | to page 132 |

The Douglas World Cruiser DWC-2 *Chicago* was fitted with floats for part of its journey. Here, a crane hoists it in Hong Kong on June 9, 1924. The *Chicago* is in the museum's collection.

CHICAGO

Carl Mitman was one of the leading curators in the Smithsonian between the 1920s and the early 1950s. He was the head curator of the Department of Arts and Industries in 1932–38 and of the Department of Engineering and Industries in 1938–1948. After World War II, as the assistant to the Secretary for the National Air Museum (founded by legislation in 1946), he effectively was the first director of the later National Air and Space Museum.

CARL W. MITMAN

ON THE SMITHSONIAN AERIAL COLLECTIONS

He wrote this June 1922 memo to William de Chastignier Ravenel, assistant to Smithsonian Secretary Abbot as curator of the National Museum's Divisions of Mineral and Mechanical Technology and proposed the reorganization of the aeronautical collections, which did occur in 1931.

I wish to bring to your attention several things which I believe, if properly carried out, will be of material benefit both to the Museum's development and consequently to the public. Upon assuming the curatorship of the Division of Mechanical Technology I fell heir to a variety of objects, some of which are wholly foreign to my experience and training, and, too, subjects which I could not become proficient in were I to devote my whole time in studying. It has been my hope that I could procure associates proficient in these standards unknown to me and thereby maintain the high standards of the Institution. . . .

There is another phase of the Museum's activities in which three divisions are involved. I refer to the subject of aerial navigation. There are involved in this subject the Division of Mechanical Technology, the Division of History, and the Division of Textiles. Surely the subject is of significant importance to be administered by a single one of these divisions; and I believe that you will agree with me that aerial navigation is purely an application of mechanical engineering and as such, properly belongs with the Division of Mechanical Technology. . . .

From the layman's point of view the manner of handling the subject of aviation by the Museum is most aggravating, for the layman, upon inquiry, is referred first to one and then to another of the divisions, and to obtain the information he is interested in obtaining, he is shunted from one representative to the other, and in nine cases out of ten never does meet the person who really could fulfill his desires.

In conclusion, therefore, I would respectfully request that you take under advisement . . . concentration of the work in the subject of Aerial Navigation under one head. In Mr. Garber, the Division of Mechanical Technology has a man who has spent a number of years on this subject and is, in addition, an experienced aviator himself, and has studied the theories of aviation a sufficient length of time to permit him to use the title of Aeronautical engineer. Having such a man associated with the Division of Mechanical Technology, it would seem logical that the whole subject of aviation should be placed under the authority of this division.

| from page 128 | Garber contacted the father of another member of his model airplane club, Admiral Strauss, the chief of the Navy's Division of Yards and Docks. With his remarkable expertise in friendly persuasion, Garber enlisted the admiral's aid in storing the wings at the torpedo factory in Alexandria, Virginia; the engines and propellers in Norfolk; and the fuselage in the Aircraft Building on the National Mall.

But for Garber's dedication and timely initiative, the NC-4 would have been forgotten and lost. Instead, in 1969, on the 50th anniversary of the first transatlantic flight, it was restored and placed on temporary display on the mall. It was then lent to the National Museum of Naval Aviation in Pensacola, Florida, where it will reside until recalled.

Important though the acquisition of the NC-4 was, Paul Garber outdid himself the following year and single-handedly made the Smithsonian a world-class aviation museum. Following the 1919 transatlantic successes of the U.S. Navy crew of the NC-4 and the British pilots John Alcock and Arthur Brown in a Vickers Vimy bomber, New York hotel owner Raymond Orteig offered a $25,000 prize to the first aviators to fly nonstop between New York City and Paris.

A Frenchman by birth, Orteig hoped to encourage improved communications between the United States and his native country. Not until 1926 did the state of aeronautical technology make this dream possible. Numerous famous and not-so-famous pilots began to build aircraft, most of them powered by reliable radial air-cooled engines to fly the 3,610 miles safely between these two great cities.

Aviation artist Geoffrey Watson evokes the landing of the Curtiss NC-4 on rough waters in this 1919 painting of the World War I flying boat.

4

One entrant was an unknown 25-year-old pilot who carried the mail between St. Louis and Chicago for the Robertson Aircraft Corporation, Charles A. Lindbergh. He felt that the way to victory was with a small, single-engine aircraft. Most of the competitors, including famed Arctic explorer Cmdr. Richard Byrd, favored large trimotor designs. Famed French ace René Fonck had crashed his massive Sikorsky S-35 on a test flight in September 1926, killing two of the four men on board, although Fonck survived. Lindbergh correctly saw that Fonck failed because of the excessive weight and complexity of the Sikorsky. For Lindbergh, simplicity meant reliability, and reliability meant success. Assembling a team of financial backers from his home in St. Louis, he succeeded in raising enough money to build his own aircraft.

After a difficult search, Lindbergh found Ryan Airlines in San Diego, whose chief engineer, Donald Hall, agreed to build a small single-engine monoplane within two months and have it ready by April 1927. Lindbergh oversaw the construction of the Ryan NYP *Spirit of St. Louis,* which was named in honor of his financial supporters. After a record-setting transcontinental flight and much additional preparation once he arrived in New York, Lindbergh took off from Roosevelt Field, Long Island, on May 20, 1927, alone, and headed for France. The country held its breath, anxious for news of his progress.

In Washington, Paul Garber was closely following Lindbergh's progress. That morning after breakfast, he composed a cablegram for Acting Secretary Charles Abbot to send

The Curtiss NC-4 rests in its beaching cradle on the shore of the Naval Air Station Rockaway in Fort Tilden, New York, before departing on its transatlantic flight.

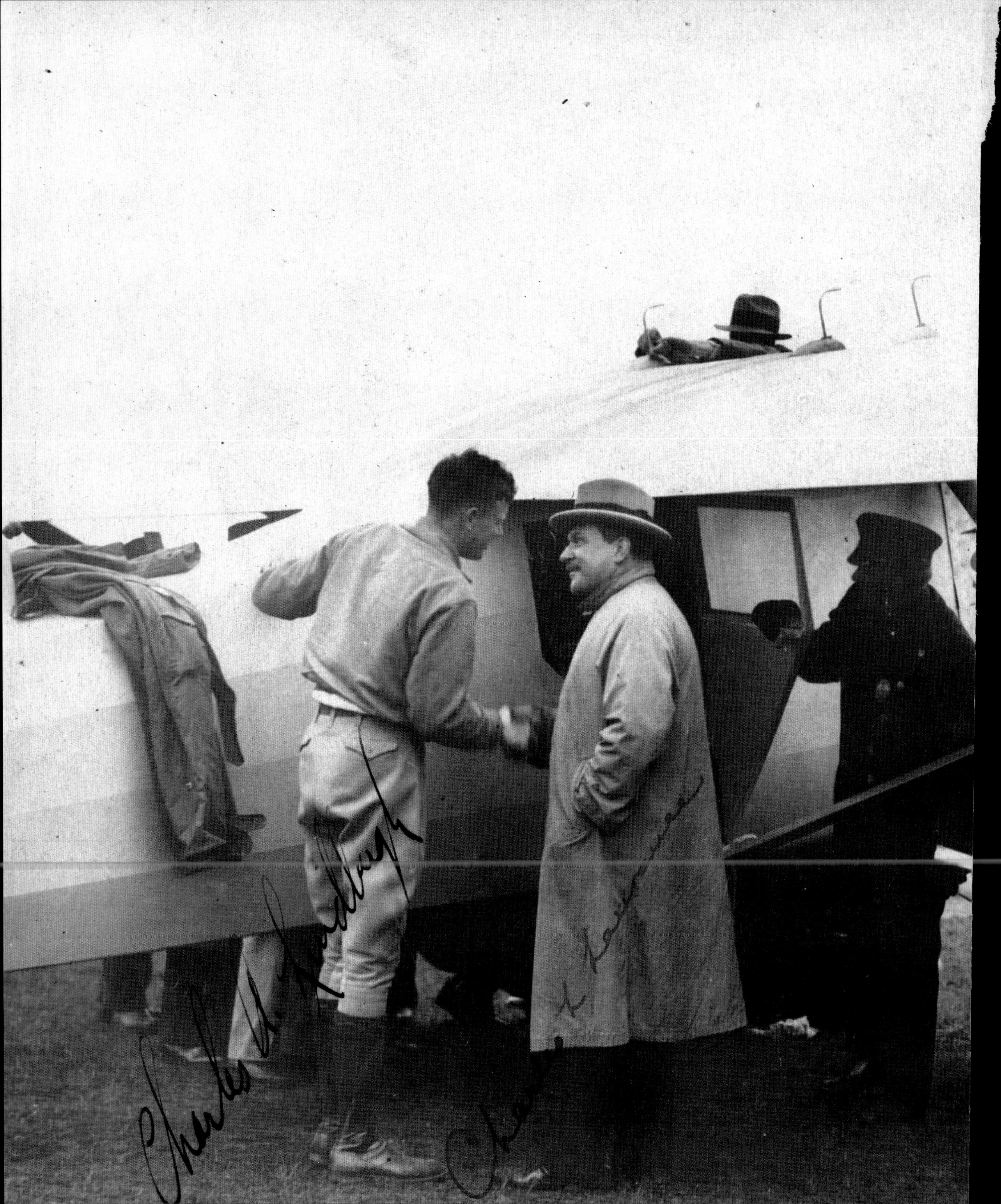
Charles A. Lindbergh

Charles Lawrance (second from left)—whose original J-1 radial engine design led to the Wright J5C that would power the *Spirit of St. Louis*—shakes Charles Lindbergh's hand moments before Lindbergh takes off for Paris on the morning of May 20, 1927.

to Lindbergh, requesting the *Spirit of St. Louis* for the national collection. When he arrived at work, Garber immediately proceeded to the Secretary's office, after convincing Carl Mitman that he should do so, to plead for his signature and approval. Garber spoke eloquently for 30 minutes, but he had difficulty persuading his boss. Abbot reminded young Garber that Lindbergh was still over New England and had a long way to go. Undaunted, Garber persisted: "I kept saying, 'We've got to get that plane, sir. He's going to make it. He's got a great plane with a great engine.'" He did not know it when he left the office, but the Secretary did indeed send the cablegram. It arrived on May 23, two days after Lindbergh completed his 33½-hour flight. It was one of the first messages he read after he woke up at the American Embassy following a much needed sleep.

The subsequent correspondence between Lindbergh and the Smithsonian is preserved in the registrar's files of the National Air and Space Museum. It was the beginning of a lifelong acquaintance between Paul Garber and Charles Lindbergh. Lindbergh was willing to give the *Spirit* to the Institution but felt obliged to check with his backers in St. Louis, as they owned the aircraft. They were so pleased with Lindbergh that they gave him the aircraft, and he in turn gave it to the Smithsonian. Actually, he sold it to the Smithsonian for one dollar. Less than a year later, on April 30, 1928, after lengthy U.S. and Central American–Caribbean tours, Lindbergh flew the *Spirit of St. Louis* to Bolling Field in Anacostia, a section of Washington, D.C., and handed it over to the Smithsonian. It has been in the Smithsonian's hands ever since, thanks to Paul Garber.

ACQUIRING THE *SPIRIT*

The story of the acquisition of the history-making airplane is much more complicated. It took a lot of polite give and take between Lindbergh and his backers before the aircraft was given to the Smithsonian. While his backers gave him the aircraft, Lindbergh insisted that their vital contribution be recognized, so their names are also on the transfer documents. Of note is the fact that the aircraft was sold to the Smithsonian.

According to the founding legislation of the Smithsonian, the Institution was a scientific establishment, and not a museum. Through the active insistence of the federal government at the turn of the last century, the Smithsonian agreed to accept the large collection of objects that the government had been acquiring, but only if the government paid for their care and protection. Under this arrangement, the U.S. National Museum was created. The museum, which was administered by the Smithsonian, was responsible for collecting, preserving, and exhibiting artifacts in the national interest. Thus, technically, the Smithsonian could not accept the donation of the *Spirit of St. Louis.* Because of this, it took another year or so for the aircraft to be transferred from the Smithsonian to the U.S. National Museum.

The Smithsonian took possession of the *Spirit* when it arrived at Bolling Field in Washington, D.C., in April 1928. With surprisingly little fanfare for such a significant artifact, the Smithsonian sent a truck and crew to the air base to remove the wings, hitch up the fuselage, and then tow this national treasure unceremoniously through the streets of the nation's capital. In so doing, the wheel bearings actually overheated and caught fire, fortunately with no damage. According to Paul Garber:

And then [Lindbergh] called me up, on April 30, and said that he was flying it in. And I said, "Well, I'll meet you down at Bolling Field."

So I did and there he flew it in and I received it. He said, "Well, here it is," and I said, "Well, thanks very much," or words to that effect. And then with the help of Sargeant [sic] Hooe and a party from the Army Air Service [Corps], we dismantled it and brought it here, hauling it up on its own wheels with the tail skid in a truck and the wings in a truck . . . and then brought it in the East Hall and there we had to take it off its wheels because the splay of the wheels was wider than we could get into the hall. . . . We brought the "Spirit of St. Louis" in and again I worked with the crew in assembling it and I went aloft in the hall and connected the cables and the pulley tackles. That was in the early days of Dr. Abbot's secretaryship because we have some pictures of him there on the floor with Sargeant Hooe and I think Sargeant Hooe has ahold of one end of the cowling and Secretary Abbot has ahold of the other end. . . .

OPPOSITE: Capt. Eddie Rickenbacker wore this coat during World War I, and Emory Bronte wore these goggles when he and Ernie Smith flew their Travel Air 5000 from Oakland, California, to Hawaii on July 14–15, 1927. The helmet's owner is unknown.

PULITZER TROPHY EARLY 1920s Sterling silver prize for winners of a national air race

Concerned that the ravages of time might cause the colors of the flags and insignias on the cowling of the *Spirit* to fade, Garber had his technicians apply a coat of clear varnish over the markings. However, the clear varnish soon aged to a honey brown color. Unfortunately, countless models, paintings, replicas and other renditions of the *Spirit* now portray the aircraft with a golden nose, a color it never had when it flew. The next time the aircraft is conserved, the museum will have a painting conservator remove the varnish. Ironically, in the 20 years during which the *Spirit* was displayed in the Milestones of Flight hall under the building's original plastic skylights, the top part of the cowling was bleached of the gold color by the sunlight. The new glass ceiling has since halted that process.

Charles Lindbergh intended that his aircraft remain forever on display at the Smithsonian, not to be removed. The Smithsonian exhibited the aircraft from the ceiling in the main hall of the Arts and Industries Building and patiently responded to hundreds of requests from museums, air shows, exhibitions, and the like from around the world begging that an exception be made and that they be allowed to display it. Weary of explaining, the Smithsonian asked Lindbergh to write a letter explaining his requirement in no uncertain terms. Lindbergh, who remained a good friend of the museum to his dying day, graciously agreed. That letter is still used today to turn down requests.

Lindbergh asked only that the door to the aircraft remain open so that visitors could see inside the cockpit and better appreciate the difficult conditions of his famous flight. For many years this was done, but it let in excessive amounts of dust. After the aircraft was conserved in 1992, the same year that Paul Garber died, a clear Plexiglas panel was affixed to the opening, which solved the dust problem.

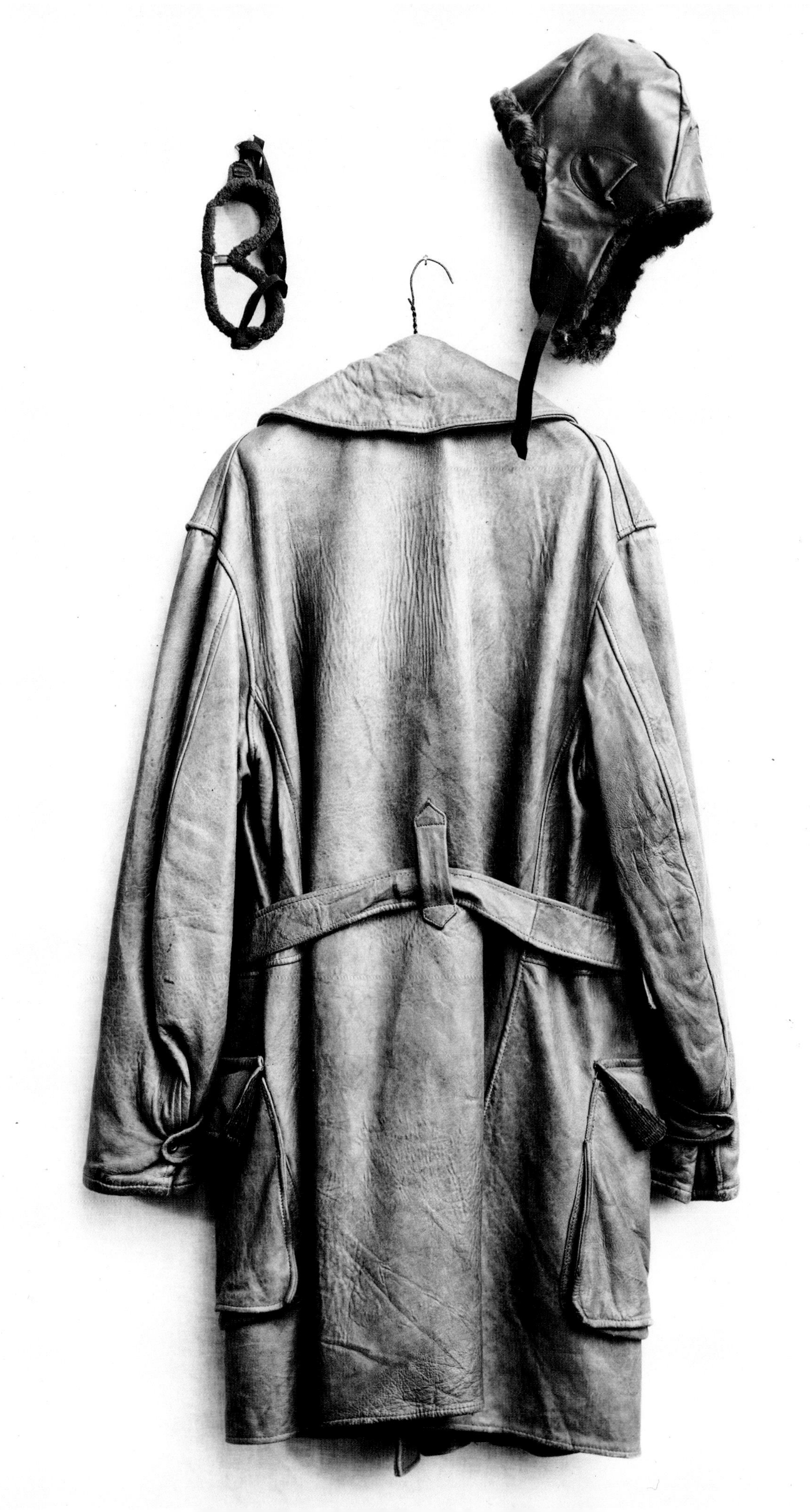

The arrival of the *Spirit of St. Louis* marked a significant milestone in the history of the museum, for its installation in the Arts and Industries Building on May 13, 1928, focused the nation's attention on the Smithsonian and its aeronautical collection. Tellingly, the public made it very clear to Smithsonian senior management that it was now intensely interested in aviation. On Easter Monday, three weeks before the *Spirit* was installed in the museum, a record-setting crowd of 1,348 people visited the Aircraft Building nearby. According to the division notes,

This is the largest single day's attendance since the Aircraft Building was opened and exceeds by 50 per cent the record established on Easter Monday of last year. There seemed to be a general impression on the part of the visitors that all of the medals and gifts bestowed upon Lindbergh were on deposit here. Everything possible was done to correct this erroneous idea. The collections, however, do include a small model of Lindbergh's "Spirit of St. Louis," made recently in the Division's Shop.

Before this event, the Smithsonian had been gathering these artifacts with no clear goal. The arrival of the *Spirit* forced the Smithsonian to acknowledge that it possessed a significant aircraft collection deserving of recognition and care. While technically a curator responsible for a myriad of technologies, Paul Garber found himself, to his delight, completely absorbed with his aircraft. According to the Division of Mineral and Mechanical Technology third quarter report for 1927,

The aftermath of the Lindbergh transatlantic flight and the aroused public

interest in aviation resulted in the call of the Division's assistant curator's whole attention, preventing his taking over any of his regular Museum work.

This included giving Lindbergh and Chief Justice William Howard Taft, head of the Smithsonian Board of Regents, a tour of the collection in December 1927. Lindbergh had just received the Smithsonian's highest honor, the Langley Medal, and he wanted to see the collection. Happily, Garber obliged, showing his distinguished visitors the future location of the *Spirit* in the North Hall of the Arts and Industries Building. This was the first of many visits Lindbergh made to the Smithsonian during his lifetime.

While the Smithsonian awaited the arrival of the *Spirit of St. Louis,* the Institution continued to expand its collection. In the summer of 1927 Acting Secretary Charles Abbot wrote to the War Department on behalf of Garber requesting the transfer of one of the famous Loening OA-1A amphibians that had completed a daring long-range group flight to Latin America. Dubbed the Goodwill Flight, these five aircraft with ten pilots on board flew throughout South America. In spite of a tragic in-flight collision that destroyed two of the aircraft and killed two of the pilots, the remaining three aircraft completed the 22,000-mile, five-month flight. The acting Secretary specifically requested one of the aircraft, the *San Francisco,* for the museum's collection.

Abbot was initially rebuffed by Assistant Secretary of War F. Trubee Davison, who said the Army still needed it, but the War Department quickly changed its mind when a promotional opportunity arose. The surviving crew members were scheduled to assemble in Washington, D.C., on December 21, 1927, the first anniversary of the beginning of the flight, to receive the MacKay Trophy and Distinguished Flying Crosses. The War Department realized that it could greatly enhance its publicity if it transferred the *San Francisco* as part of the ceremonies. The attention would benefit the museum as well. Curator Carl Mitman announced:

On December 21st, the Distinguished Flying Cross will be presented to the Army aviators who made the flight last year around South America, at a luncheon in the Pan American Union Building, to which all the Latin American diplomats and the Secretary of State have been invited. Following the luncheon the distinguished company will come to the Aircraft Building.

It goes without saying that news photographers will want to take pictures of this assemblage. It will be quite a publicity stunt for us and I feel that we should do everything in our power to aid.

In exchange, the museum was willing to give up the Martin bomber that had occupied a corner of the building for several years. In a remarkably nearsighted statement, Assistant Secretary Abbot asserted, "This obsolete machine has no distinctive history or record and is in rather bad condition." Harsh words for the first American-designed strategic bomber. While the acquisition of the Loening was a major triumph, the loss of the Martin was an unfortunate mistake, since it was eventually destroyed by the Army. None of these bombers remain.

LINDBERGH'S PRIZE 1927
Illuminated check for $25,000 presented to Lindbergh for his successful non-stop trans-atlantic flight

OPPOSITE, TOP: During his visit to the Panama Canal Zone in January 1928, Lindbergh flew aerobatics in a borrowed Army fighter over Gatun Lake.

OPPOSITE, CENTER: Lindbergh's Ryan NYP *Spirit of St. Louis* was installed in the North Hall of the Smithsonian Institution's Arts and Industries Building on May 13, 1928. Smithsonian Secretary Charles G. Abbot (right) and the Army sergeant in charge of the project pose holding the aircraft's engine cowling.

OPPOSITE, BOTTOM: *Spirit of St. Louis* was suspended just above the exhibit cases in the North Hall.

Paul Garber and the Smithsonian had engineered two coups in acquiring the Loening and especially the *Spirit of St. Louis.* The nation was becoming "air minded," as the so-called Lindbergh boom of interest in all things aviation swept the nation. But while the nation was focusing its aviation interest on the Smithsonian's collection in unprecedented intensity, tight economic times continued, despite the rapidly expanding economy. Garber's reward for his stellar work over the previous two years was to receive a six-month furlough beginning on May 1, 1929. As unfair as it may have looked, others in the division had taken their furloughs in turn in order to stretch the thin federal appropriation as far as possible. He was temporarily replaced so that the Institution could hire a laboratory mechanic to deal with a backlog of maintenance work. Garber returned on November 1.

The arrival of Lindbergh's *Spirit of St. Louis* also focused the attention of the Smithsonian on its growing collection of aeronautical artifacts. Hitherto, the Castle—as insiders called the Smithsonian's administration and administrators—paid little heed to this collection. Science was its primary interest. Garber persisted, gathering aircraft together with other mechanical objects when available; but the administration had no collecting plan or any sense that aircraft constituted a separate collection. This state of affairs changed when the Castle was confronted by the record numbers of visitors who sought out the *Spirit* and other aeronautical objects. Garber recalled many years later that acquiring the *Spirit of St. Louis* was "the turning point," and it "gave a much greater breadth to the subject of aeronautics in this institution." He continued, "With that accession, aeronautics became recognized and within a short time a Section of Aeronautics within the Division of Mineral and Mechanical Technology was established and I was favored to be put in charge of it."

REORGANIZATION

In 1931, despite the fact that the nation was in the depths of the Great Depression now that banks had collapsed globally, the Smithsonian reorganized its management to reflect the growing importance of the aviation collection. On July 18, with the approval of the Civil Service Commission and the Personnel Classification Board, the Division of Mineral and Mechanical Technology was reorganized and renamed the Division of Engineering within the Department of Arts and Industries. The new Division of Engineering now consisted of three sections: the Section of Mechanical Technology, under assistant curator Frank A. Taylor; the Section of Mineral Technology, under curator Carl W. Mitman, who was also the head curator of the division; and, the Section of Aeronautics under Paul E. Garber, who as assistant curator was now given direct curatorial responsibility over the aviation collection. One year later, the position of director of the Department of Arts and Industries was abolished and replaced with the new post of head curator. Carl Mitman was promoted to this slot, and Frank Taylor moved up to become curator of the Division of Engineering.

OPPOSITE, TOP: On July 20, 1934, Wiley Post crashed his Lockheed Vega *Winnie Mae* in Flat, Alaska, having flown directly from Siberia during his second around-the-world flight. The plane was quickly repaired, and Post completed his flight in record time.

OPPOSITE, BOTTOM: *Winnie Mae* on exhibit in the Arts and Industries Building

FIRST PRESSURE SUIT 1934–35 Invented by Wiley Post, working with the B.F. Goodrich Company

Mitman was an excellent curator, and Paul Garber admired him greatly. But they did have their differences concerning the importance of aviation and Garber's penchant for collecting—differences that were often aired in energetic discussions. For example, Garber argued to acquire the Douglas World Cruiser so it could be preserved, but he did not seem to be able to persuade Mitman.

Mitman was a mining engineer, and although I had the highest reverence and affection for him because I sort of loved the man, yet, sometimes when he turned down my suggestions, as he did the time that I wanted the B-10 that had flown to Alaska, I remember that he gave me a rather complete dressing down that I was getting too many airplanes. I've been told that quite often, that we had no space . . . for them, and I'd have to cram things together to get one more in. I was told one time that the museum was for people too, not just for airplanes alone. We had to somehow fit these, our customers, in there some way. But I remember sort of snorting to myself, not out loud, but saying to myself, "There [is] nothing farther apart than a mining engineer and an aviator: One's way underground, the other's up in the sky."

The reorganization also coincided with the revamping of the Aircraft Building, which, after several years of work, reopened with a new permanent collection. The building was fitted with a sprinkler system and other upgrades that closed the structure from October 7, 1930, until June 14, 1931. When it reopened, visitors found a new aircraft on display, the *Bremen,* the first aircraft to fly nonstop across the Atlantic from east to west. The historic Junkers W33, which had crossed the ocean with its three-man crew on April 12–13, 1928, was borrowed from the Museum of Science and Industry in New York City because that museum lacked exhibition space.

In a dramatic acquisition, Harold Pitcairn and his Autogiro Company of America presented to the Smithsonian the first autogiro to fly in America, landing the aircraft on the National Mall in front of the Arts and Industries Building. Piloted by James G. Ray, the autogiro first flew in 1928 and was the prototype for a series of short takeoff and landing aircraft, precursors to the helicopter. After Secretary Charles Abbot accepted the aircraft, the technicians dismantled it and wheeled it straight into the Arts and Industries Building, where it was quickly reassembled and placed on display.

A steady stream of engines, propellers, accessories, and models were acquired throughout the period, but space constraints were already negatively affecting the museum. In a complaint that has echoed repeatedly throughout the decades, Paul Garber feared that a lack of space was hurting the collection. In his fiscal year 1934 annual report, he claimed that while the Aircraft Building was essential, it was too small:

Its fifteen years of service as an aircraft museum show a constant struggle to protect this "temporary" building against

THE WINNIE MAE OF OKLAHOMA
LOS ANGELES TO CHICAGO · 9 HRS. 9 MIN. 4 SEC. · AUG. 27, 1930
AROUND THE WORLD · 8 DAYS, 15 HRS. 51 MIN. · JUNE 23 TO JULY 1, 1931

NR105W
LOCKHEED
THE WINNIE MAE OF OKLAHOMA
LOS ANGELES TO CHICAGO · 9 HRS. 9 MIN. 4 SEC. · AUG. 27, 1930
AROUND THE WORLD · 8 DAYS, 15 HRS. 51 MIN. · JUNE 23 TO JULY 1, 1931
AROUND THE WORLD · 7 DAYS, 18 HRS. 49 MIN. · JULY 15 TO JULY 22, 1933
LOS ANGELES TO CLEVELAND IN THE SUB-STRATOSPHERE. 340 M.P.H. MAR. 15, 1935

the elements and to ingeniously adjust its contents so as to take advantage of every bit of space, so that a maximum number of examples of this important period of development could be preserved. The saturation point is, however, fast approaching and lack of space presents an increasing problem in the matter of new exhibits. Several possible accessions were lost to the Museum during the year, and improvements in present exhibits have had to be curtailed. Additional aircraft can be accepted only through the retirement or loss of present displays, or by using the air space above the exhibits of other divisions in the Arts and Industries Building, a practice which is discouraged.

It is hoped that the continued careful placement of specimens will result in a pleasing appearance of the exhibition, and perhaps some new source of space may become available.

In August 1935, famed aviator Wiley Post and humorist Will Rogers died in a plane crash while in Alaska. Seizing this tragedy as an opportunity to gain public support for a bigger museum, Garber's immediate supervisor, Frank Taylor, prepared a detailed proposal for a Will Rogers memorial aircraft museum in the nation's capital. Taylor recommended that the existing national collection be used as the core of the new Smithsonian museum and that the existing Aircraft Building be replaced by a larger, permanent structure suitable for a national collection. The Rogers memorial museum would occupy 55,000 square feet, some 20,000 square feet more than the existing building. Taylor and Garber estimated the cost for two "monumental-type" and expandable buildings | to page 148 |

Herbert Hollick-Kenyon (standing in cockpit), the pilot on Lincoln Ellsworth's 1935 trans-Antarctic flight, shakes hands with Paul Garber, who accepts the aircraft for the Smithsonian Institution.

POLAR STAR
TEXACO

Amelia Earhart, Wiley Post, and Roscoe Turner (left to right) inspect a radial aircraft engine at the Lockheed Aircraft Corporation factory in Burbank, California, July 18, 1935.

| from page 144 | at $720,000, with an additional $100,000 set aside to purchase future artifacts. According to Taylor:

The National Aircraft Collection is at present a living institution. The buildings in which it is housed are visited annually by over one million persons. Its functions are educational and historical. It has been aptly phrased a Valhalla of aeronautics to which worthy planes go when their days are over to keep alive the memory of their accomplishments and those of the men who built and flew them. The inspirational value of the collection cannot, of course, be measured but it must be tremendous. It is difficult to imagine a more fitting national memorial to Will Rogers than a building to house this collection.

This farsighted proposal, the first for a separate national air museum, failed to get enough support. It was, however, an important first step.

Regardless, Garber and the Section of Aeronautics soldiered on, collecting significant aircraft whenever possible while rearranging the existing collection to make the new acquisitions fit. In 1934, he has able to bring in the Wright EX biplane, the *Vin Fiz,* which was the first aircraft to fly across the U.S.

Flown by Calbraith Perry Rodgers in 1911, the aircraft left Sheepshead Bay, Long Island on September 17 and arrived in Pasadena, California, 4,231 miles later, on November 5, the official end of the flight. (Rodgers later flew the aircraft to the coast and dipped the *Vin Fiz's* wheels in the waters of the Pacific Ocean.)

While the total flight time was 3 days, 10 hours, and 4 minutes, the

The Douglas DC-3 revolutionized commercial air travel because of its efficiency and productivity. It entered service in 1936, and thousands flew for decades, including this DC-3 of Eastern Air Lines at Boston's Logan Airport in 1948. In the background are a red Cessna Bobcat of New England Central Airlines and a four-engine DC-4.

THE GREAT SILVER FLEET
EASTERN
370
NC28382
U.S. AIR MAIL
AIR EXPRESS

Adaptable Annie

TOP: Throughout his life Paul Garber loved aeronautics in all of its forms, and he maintained a wonderful sense of humor. He is shown here with the helmeted Benjamin Lawless in a balloon on the National Mall on April Fools' Day 1966.

BOTTOM: Garber (left) and Capt. Charles E. "Chuck" Yeager inspect the rocket engine of the Bell X-1 *Glamorous Glennis* at Andrews Air Force Base, Maryland, August 28, 1950, before it was delivered to the Smithsonian.

OPPOSITE: Paul Garber stands next to the Boeing 247-D *Adaptable Annie* after it was transferred to the museum in 1953. It was the first 247-D built. Roscoe Turner flew it in the MacRobertson Race from England to Australia in 1934, and then it was delivered to United Air Lines.

actual elapsed time was 49 days, during which Perry made 69 landings, many of which damaged the aircraft. Most of the aircraft's parts had been replaced by the time it reached California. When Garber acquired the Wright brothers' transcontinental flier from the Carnegie Museum in Pittsburgh, it was in poor shape. Garber "saw that poor thing quite neglected and just crying for some tender loving care, and I finally got it and I, personally, physically repaired it and put it on display."

In 1936, Garber collected the *Polar Star,* a Northrop Gamma that had completed the first crossing of the Antarctic continent—a distance of 2,400 miles from Dundee Island to within 16 miles of Little America. Famed explorer Lincoln Ellsworth had purchased the *Polar Star* for flights in Antarctica. In 1935, the aircraft's range and flexibility enabled him and pilot Herbert Hollick-Kenyon to cross unexplored territory, which he named James W. Ellsworth Land, after his father. Ellsworth gave the aircraft to the museum with the understanding that since it was still a relatively new aircraft, he could recall it for another flight if he wished. Fortunately for the museum, that time never came.

The Gamma was designed by John K. "Jack" Northrop, one of the greatest aircraft designers in history, who had earlier designed the Lockheed Vega series of aircraft and the multicellular wing—which gave the Douglas DC-3, perhaps the greatest transport aircraft in history, its phenomenal strength. He also pioneered all-metal aircraft construction and the flying wing in America.

Yet the museum did not thoroughly appreciate | *to page 156* |

The Curtiss R3C-2 Racer hangs proudly in the Barron Hilton Pioneers of Flight gallery.

Menacingly beautiful with its black fuselage and gold-colored wings, the Curtiss R3C dominated the Pulitzer Trophy and Schneider Cup races in 1925, setting an absolute speed record along the way.

CURTISS R3C-2

WORLD'S FASTEST PLANE IN 1925

Early on in aviation's development, air races began to enjoy a worldwide popularity. Two of the most coveted prizes were the Schneider Cup and the Pulitzer Trophy, each funded by an individual. In 1912 a wealthy French aviation enthusiast, Jacques Schneider, established an annual trophy for an overwater race by seaplanes. At about the same time an American newspaperman, Ralph Pulitzer, sponsored the Pulitzer Trophy Race to promote high speed in land planes.

In 1925 the U.S. Army and Navy ordered similar R3C racers from the Curtiss Aeroplane and Motor Company. Powered by a sleek Curtiss V-1400 12-cylinder engine that generated 610 horsepower using a noxious mixture of benzol and gasoline, the R3Cs ran away with first place in both trophy races that year.

The land-based version, known as the R3C-1 and piloted by Army Lt. Cyrus Bettis, won the Pulitzer Trophy Race on October 12, 1925, at Mitchel

Field, Long Island, at a speed of 248.9 miles an hour. On October 25, fitted with streamlined floats and redesignated the R3C-2, the aircraft was piloted to victory by Army Lt. James H. "Jimmy" Doolittle in the Schneider Cup Race outside Baltimore. His average speed was 232.57 miles an hour.

The day after, Doolittle flew the R3C-2 on a straight course at a world-record speed for seaplanes of 245.7 miles an hour. (Doolittle would later dominate the National Air Races, earn a doctorate from MIT, help develop 100-octane fuel, lead the famous Tokyo Raid in April 1942, and command the Eighth Air Force in Europe.)

In the Schneider Cup Race of 1926, this same airplane, piloted by Marine Lt. Christian F. Schilt, USMC, and powered by an improved engine, won second place with an average speed of 231.4 miles an hour.

Pilot Lt. James H. "Jimmy" Doolittle stands on the starboard float of the 1925 Schneider Cup–winning Curtiss R3C-2.

| from page 151 | the design of the aircraft, nor its designer. Head curator Carl Mitman, with the concurrence of Paul Garber, was willing to accept the Gamma because it represented a new generation of cantilevered (or internally braced), low-wing all-metal aircraft—but only if the *Bremen* it would replace could find a good home. Writing to Assistant Secretary J. E. Graf, Mitman claimed that the *Bremen,* "which is similar in design to the Northrop, is the more fundamental type. Furthermore it is the product of Junkers, one of the foremost aeronautical engineers in the world. The Northrop, on the other hand, is the product of Jack Northrop, a far less prominent designer and not possessing the world-wide recognition of Junkers."

Fortunately, the *Bremen* found a good home in the Henry Ford Museum, and Jack Northrop was eventually recognized as one of the all-time great aircraft designers.

Perhaps Garber's most difficult acquisition came in 1937, when he brought Wiley Post's Lockheed Vega, the *Winnie Mae,* into the collection. Post's death in 1935 with Rogers, in a crash of their modified Lockheed, left Post's widow, Mae Laine, destitute. Much as she wished to give the *Winnie Mae* to the nation, Post could not afford to do so. Using his growing political connections, Garber persuaded Congress to pass a $25,000 appropriation for the purchase of the now famous aircraft and personally arranged for the transfer of the aircraft to the museum.

He spent weeks in Oklahoma preparing, packing, and shipping the *Winnie Mae* and Wiley Post's most significant artifacts, one of which was his pressure suit, the first of its kind. The *Winnie Mae* was of advanced design, but it was constructed of plywood. Although it was strong, the fuselage could not withstand the stress of pressurization at high altitudes. The obvious solution was to find a way to protect just the pilot in a pressurized suit and helmet. In creating the grandfather of all spacesuits, Wiley Post worked closely with the B.F. Goodrich Rubber Company to develop a garment that would protect him from hypoxia while flying in the stratosphere.

OPPOSITE: The Aquabelles from the New York World's Fair of 1939 and Army Air Corps officers pose on the wing of Boeing's experimental XB-15 bomber.

THOMPSON TROPHY 1930 Won by Roscoe Turner during a closed-course competition at the National Air Races

Although it took several iterations, the third version, complete with its diving-style helmet, enabled Post to reach an unofficial altitude record of over 50,000 feet in 1935. While flying above 35,000 feet, Post discovered a "river of air"—the jet stream—which pushed his 170-miles-an-hour Lockheed Vega across the continent at 340 miles an hour. Thus, in addition to setting records, Post made two extremely important contributions to flight, both of which are commemorated by his collection.

The method of acquisition of the *Winnie Mae* was unique. Traditionally, the Smithsonian does not pay for artifacts. Despite its seemingly large budget, the Institution simply does not have enough resources to do so, with the exception of the occasional one-dollar purchase for legal reasons of objects like the *Spirit of St. Louis.* Mae Laine Post was truly desperate. Her husband's untimely death left her with little except his aircraft and personal collection.

Fortunately, Oklahoma Representative Joshua Bryan Lee had earlier tried to get Congress to reward Wiley Post with a cash gift. After Post's sudden death, Lee pressed on and secured $25,000 for Mae Laine. Paul Garber worked with Lee to ensure that the government received something for its efforts and had the legislation written to have Mae Laine donate the *Winnie Mae* and other important aeronautical objects to the Smithsonian, which she gratefully did. This greatly enhanced the national collection and gave Mae Laine a means to survive financially.

By the end of the decade, with war engulfing Europe and Asia, and the United States becoming increasingly involved, the problems of the Smithsonian and its aeronautical collection took a backseat to more pressing national issues. Nevertheless, a small but persistent effort to expand the museum continued. Favorable articles appeared sporadically in local newspapers and national magazines extolling the virtues of the Smithsonian's excellent but largely overlooked aeronautical collection. In July 1940 Edward Lawson wrote in *Popular Aviation* about the "thrilling story of the aviation industry" that was on display in Washington, D.C.

"Unfortunately," he wrote, "each year thousands of air-minded visitors to the nation's capital miss this treat because they don't know it exists. They see the famous cherry trees, the White House, the Monument and the Capitol, but too frequently they pass without a second thought the low-roofed, barn-like hangar midway between the latter two, where the nation's aviation treasures are displayed."

Other concerned citizens began a quiet campaign to secure a proper building to house the aeronautical collection. For months Al Williams,

one of the best known exhibition pilots in America, had been campaigning throughout the country to raise interest in gaining the return of the 1903 Wright Flyer to the United States, to the Smithsonian in particular. While significant progress had been made in assuaging Orville Wright's concerns about the Smithsonian's interpretation of the Wrights' role in aviation, Williams was also concerned that once the Flyer came home, the existing facilities would not do justice to this milestone artifact. In a 1937 edition of the *Washington Post,* author Bob Ball argued vociferously for a new building for the Flyer and all of the national aeronautical collection:

This country boasts scores of priceless souvenirs of its remarkable progress in aviation. Jealously it guards them . . . but poorly it houses them.

In an obscure corner of the Smithsonian grounds stands (shakily) a long tin shack, constructed during the war for the testing of Liberty engines. On a rainy day its roof leaks. On hot summer days the rays of the sun are intensified by its tin roof. It is the "Air Museum."

Paul Garber, curator of aeronautics, and his staff, have gathered together historical planes, propellers, maps and photos worthy of a finer hall. . . . The need for larger quarters is imperative.

But let's ask President Roosevelt to appeal to Congress . . . or the country at large, to build for our great collection a marble hall, beautiful and permanent, commensurate of our foremost position in world aviation . . . and a great domed central hall for the finest treasure of them all, the Wright 1903 plane. | to page 163 |

A Grumman TBF Avenger with wings folded taxis down the flight deck after landing on an aircraft carrier in the Pacific Ocean during World War II. Two other aircraft are preparing to land behind the Avenger.

COOK

A stricken American battleship fights for its life after the Japanese attack on Pearl Harbor, December 7, 1941. OPPOSITE, TOP: Japanese Mitsubishi A6M2 Model 21 Zero fighters prepare for launch against Pearl Harbor. OPPOSITE, CENTER TOP: A U.S. Navy Grumman F6F Hellcat warms up before a mission in the Pacific. OPPOSITE, CENTER BOTTOM: A flight of Consolidated B-24 Liberators of the 15th Air Force bomb the Concordia Vega oil refinery, Ploesti, Romania, on May 31, 1944. OPPOSITE BOTTOM: Capt. Raymond M. Walsh in his Republic P-4 Thunderbolt is outlined against the flaming explosion of a Nazi ammunition truck that he has just destroyed in France in 1944.

23

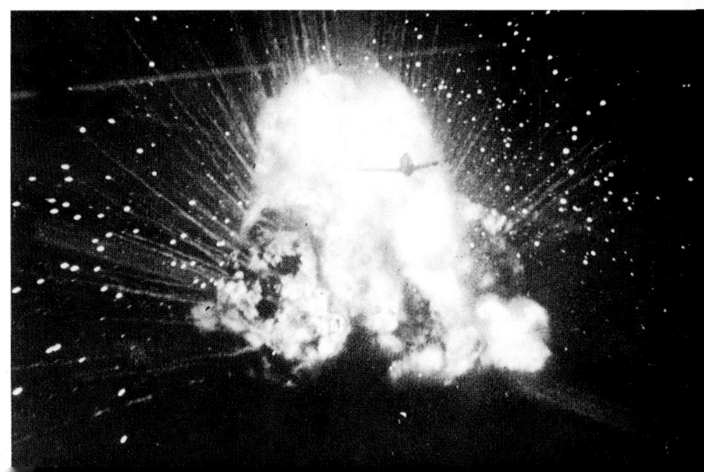

An experimental-scale, wood-and-metal wind tunnel model of a North American B-25 Mitchell bomber is being prepared for a test in 1942 at North American Aviation's factory in Inglewood, California.

| *from page 159* | During 1940, Secretary Abbot directly petitioned President Roosevelt for a tract of land near the U.S. Capitol to address these very issues. For the moment, understandably, the request had to wait as the nation was heading toward war. As the situation deteriorated in Europe, officials in the government and the Smithsonian grew increasingly concerned about protecting irreplaceable national artifacts should the war expand to North America. As a result, under executive order of President Franklin Roosevelt, the Committee on Conservation of Cultural Resources was formed in September 1941. Immediately the committee, under the leadership of the Librarian of Congress, Archibald MacLeish, with National Museum associate director John Graf as a prominent participant, began to forge plans for the protection and evacuation of national treasures.

The responsibility for drafting a protection plan for the Smithsonian artifacts fell on the capable shoulders of head curator Carl Mitman, with final approval by Secretary Abbot. Mitman appointed air-raid wardens for each Smithsonian building and had air-raid shelters constructed in the Freer Gallery and the Natural History building; in addition, first-aid stations were set up throughout the Institution.

Following the Japanese attack on Pearl Harbor on December 7, 1941, and America's subsequent entry into the war, Undersecretary Alexander Wetmore began a search in the Washington area for a suitable safe storage site. Following the recommendation of F. M. Setzler, head curator of the Department of Anthropology, the Smithsonian gained the use of a large stone-and-concrete building near the headquarters of Shenandoah National Park in Luray, Virginia.

Large, strong, invisible from the nearest highway, and 50 miles away from downtown Washington, the building was well suited for temporary storage of the Smithsonian's most valued items. Inevitable bureaucratic delays impeded the transfer, but by December 9, some 120,000 pounds of objects had been safely moved under Mitman's leadership. Both the Langley aerodrome and the Wright 1909 Military Flyer were moved. Interestingly, the *Spirit of St. Louis* remained in place, protected with the rest of the aeronautical collection in Washington for the duration of the war.

During World War II, the Smithsonian devoted much of its work to supporting the war effort and maintaining its existing collections. Active collecting was put on a back burner, although the museum was able to collect the Bell XP-59 Airacomet, America's first jet aircraft.

In 1942 the Smithsonian formally apologized to the Wrights in its annual report, recognizing the brothers' work as original, thus paving the way for the eventual return of the 1903 Flyer. Throughout the war, head curator Carl Mitman chaired a pan-Institutional committee that helped the Institution lend its scientific and technical expertise to the war effort. Once his tasks were completed, Mitman joined the Army. Paul Garber had already joined the Navy.

| *to page 167* |

BOMBER JACKET
Ca 1944
A-2 flight jacket worn by the radio operator of the Martin B-26 Marauder "Flak Bait" during World War II

Paul Garber was the first curator of the Smithsonian aeronautical collections and the founder of the National Air and Space Museum. He liked to tell the story of how his target kites were credited with saving an American carrier from enemy attack.

PAUL GARBER

ON WORLD WAR II TARGET KITES

One day the chief of the Bureau of Aeronautics, Vice Admiral DeWitt Clinton Ramsey, strode into my workshop and demanded if I was the "kite person." I stammered a quick "yes, sir," after which he put me at ease, stating: "That kite of yours may have saved a ship." He told me of a task force that was steaming in enemy waters. Aboard ship there was gunnery practice with several men in their gun tubs shooting at "my" kites. . . . The day was hazy enough to reduce visibility but that made the training more realistic, as though the Zero airplanes painted on the covers of the kites were dodging in and out of clouds. Suddenly—so close to the water that the radar had not detected them—came two Japanese torpedo bombers. Somebody grabbed a squawk horn and shouted, "Change targets from kites to enemy airplanes off the port beam!" The kite shooters swung their guns around and shot those rascals into the water. . . . From then on, if the Admiral happened to be with a group and saw me, he would ask them, "Did anyone ever hear how a kite saved a carrier?"

Paul E. Garber holds the Mark I target kite he designed for shipborne anti-aircraft practice when he was a Navy commander.

C 4988

| from page 163 | Garber's lifelong interest in model building contributed to America's war effort in a most unusual way. Throughout the Aircraft Building, he had constructed many different model aircraft and dioramas to enhance the exhibits. In 1940, he made a detailed model display of contemporary military aircraft that were currently in the news. Using photographs mounted on masonite and aircraft-plan forms cut from plywood, Garber's exhibits attracted the attention of Cmdr. Luis de Florez.

On Monday, December 8, 1941, the day after the Japanese attack on Pearl Harbor, Commander de Florez called Garber about borrowing the display for the Navy. With the Secretary's permission, Garber dismantled the two cases and delivered them to the Navy Department, where he reinstalled the cases and discussed model making for the rest of the day. Garber had made a definite impression, as the Navy desperately needed models to train a new generation of pilots and sailors in aircraft recognition.

One week later, he was introduced to the chief of the Bureau of Aeronautics, Adm. John Towers. The admiral had just received technical information and photographs from Pearl Harbor of the attacking Japanese aircraft, particularly the deadly Mitsubishi A6M Zero.

Garber took the information, incorporated his own data, and quickly produced a large 1:16 scale model of the Zero. Employing his expertise in constructing educational exhibits with models, he emphasized the importance of using a model that could be held while studying all of its angles, in three dimensions rather than two.

Towers was impressed. He instructed Garber to begin to mass-produce these models, settling on a smaller, more practical scale of 1:72. Now working almost exclusively for the Navy, Garber asked several companies to set up a production line. He eventually contracted with the Cruver Company of Chicago, whose injection-molded plastic products proved lighter than metal and more durable than rubber. Garber brought together a talented team of model builders from throughout the Navy to design and build the ever increasing number of models of friendly and enemy aircraft. Eventually, several other manufacturers established production lines that would make thousands of black plastic identification models that proved invaluable to America's war effort by helping both civilian and military personnel identify aircraft spotted from the ground or the air. Garber kept two of each model—one for his office and one for the Smithsonian.

Garber did all of this on his own time. Many weeks after he had begun to help the Navy, he received a call from the museum, informing him that he had used up all his annual and sick leave. Eager to contribute to the war effort but concerned that he could not continue, he asked if the Navy could arrange to hire him as a temporary employee. The Navy did one better: The commander in charge of the model project spoke to his superiors, and soon Garber was made a lieutenant in the Naval Reserve, assigned to the Bureau of Aeronautics, Section of Special Devices.

With his status now confirmed, Paul Garber devoted the next five years to developing better models and gunnery aids for the Navy. With the help of Cornelius Roosevelt, the grandson of Theodore Roosevelt, Garber developed a gunnery simulator, using a repeating crossbow to teach lead angles. Working from a garage at 610 H Street, N.E., Garber's team also developed cardboard aircraft models that could be shipped flat and assembled in the field. With the wood-manufacturing expertise of the Mohler Organ Company in Hagerstown, Maryland, the Section of Special Devices designed and built cockpit familiarization trainers for student pilots. Throughout this time, Paul used his model-making expertise to develop curricula in aircraft identification classes.

Garber's most notable contribution came from his first love: kites. Learning gunnery was a daunting task for the new recruits, and crews had trouble maintaining their proficiency while at sea. Garber and his friend and co-worker Stanley Potter concocted a unique solution to this problem: large maneuverable target kites. Garber and Potter devised an Eddy-type deltoid kite with a fin and rudder. Garber eagerly approached de Florez, now a captain, with his new creation, and after much persuasion, coaxed the

TARGET KITE 1943–44 Marked with the shape of the Japanese Zero fighter

OPPOSITE: Paul Garber (left) shows Col. Enrique Flores, a Chilean Air Force air attaché, a model of Lt. Dagoberto Godoy's Bristol M.1C Monoplane Scout. Godoy made the first west-to-east crossing of the Andes, from Santiago, Chile, to Mendoza, Argentina.

busy captain to the roof for a demonstration. Despite numerous telephone and power lines surrounding the roof, Garber flew the kite effortlessly, spelling out de Florez's name in the sky. Deeply impressed, Captain de Florez "whapped me on the back and said it was 'the best damned target he had ever seen.'"

Six target kites were immediately ordered, and when word spread, another 100 were ordered for the U.S.S. *Yorktown,* sewn by the women of the Navy Yard's sail loft. Business was good. After 1,000 were made, Garber was ordered on a demonstration tour of the Navy's East Coast gunnery stations. By the end of the war an astounding 300,000 Garber target kites had been built. The Navy even patented the kites in his name.

He continued his work with kites for the Navy, developing a communications kite to transfer messages from aboard ship to a passing airplane. The task required a much larger kite, capable of lifting 300 feet of heavy snatch line and a message pouch. Garber designed a large three-triangular cell box kite that worked quite well during tests but did not see widespread service.

By the end of the war Paul Garber had reached the rank of commander and was eager to return to his civilian duties at the Smithsonian. Much was happening at the museum in the immediate postwar glow of victory, and he wanted to participate. The Navy had other ideas, however, and kept him in uniform until 1946 to write a history of wartime achievements of the Section of Special Devices. When he did return, he found that much had changed.

Lindbergh's Ryan NYP *Spirit of St. Louis* is suspended from the rafters of the Arts and Industries Building. The original 1814 Fort McHenry flag—the "Star-Spangled Banner"—is mounted on the wall behind it.

Spirit of St. Louis

Posters have been a significant means of mass communication throughout their history, often with striking visual effect and powerful impact. They have been used as a way of advertising and marketing to large target audiences.

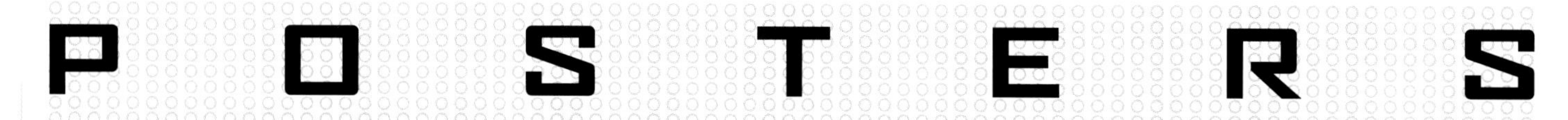

1915

1989

1953

1953

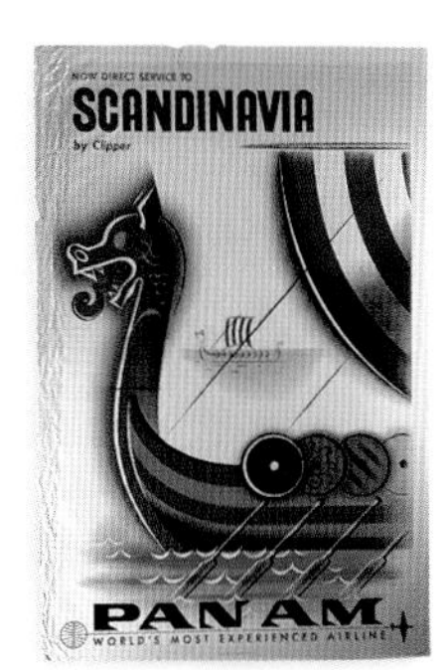

1957

1957

1938

Ca 1980

1919-1924

1951

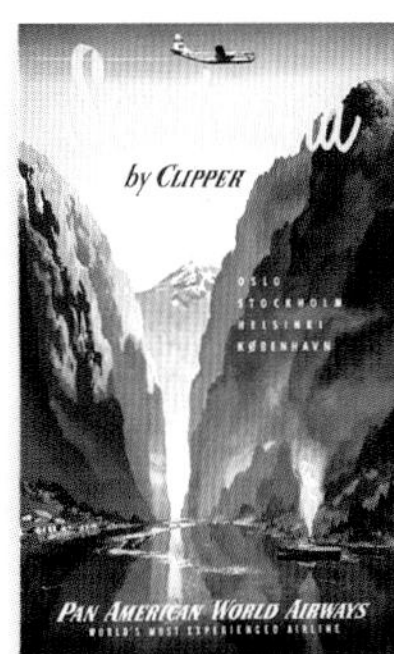

1951

1951

1951

1951

1943

1943

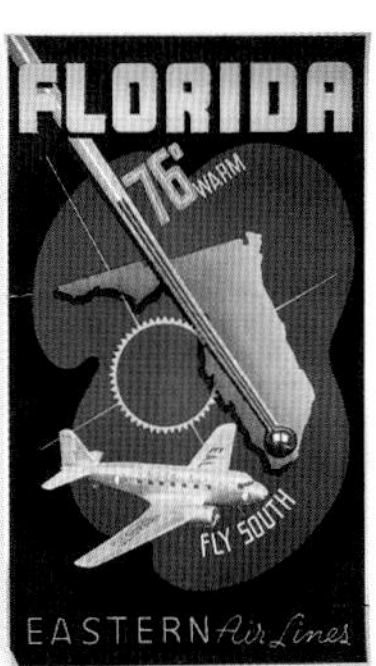

1938

1942

Ca 1963

1917-1918

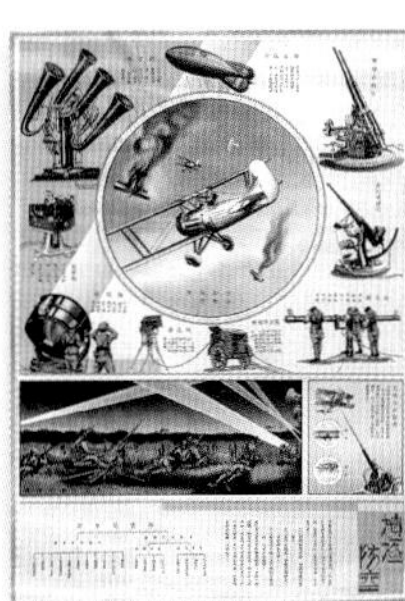
Ca 1924

1932

1939

Ca 1933

1929

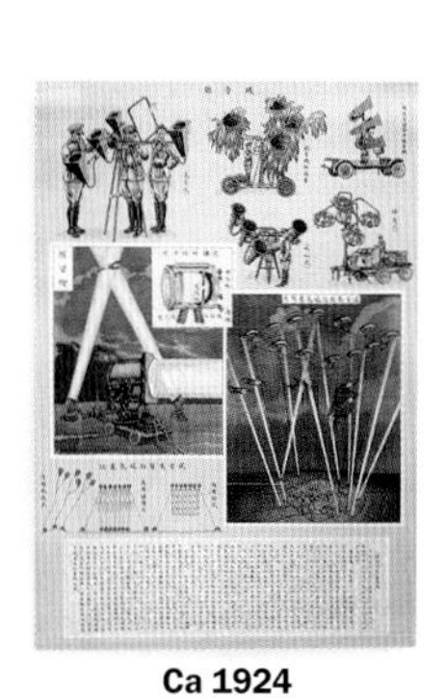
Ca 1924

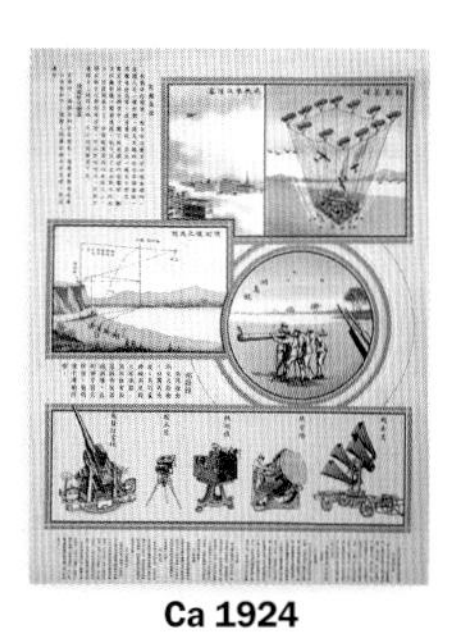
Ca 1924

Ca 1924

1932

Ca 1944

1917

1919

1917

1933

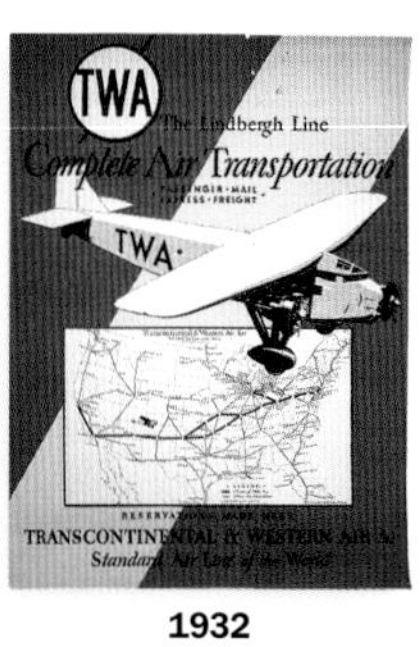

1932

1944

1932

1944

1944

1944

1944

1944

1944

1933

1930

1945

1910

1910

1946

The museum's collection of more than 1,300 posters, quite possibly the largest of its kind publicly held in the United States, focuses primarily on 19th-century ballooning exhibitions, early 20th-century airplane demonstrations and air meets, and advertising of aviation-related products and activities.

Printing technologies represented include original lithography, silk screen, photolithography, and computer-generated imagery. The collection is significant for its aesthetic value and as a unique representation of the cultural, commercial, and military history of aviation. It showcases stunning graphic design and reveals popular attitudes of the era. It also spotlights an intense interest in flight, both public and private, during a period of its technological and social development.

The poster collection constitutes a rare national treasure whose preservation is essential if we are to understand how aviation was represented and understood in the 20th century.

1945

2.5 ROBERT GODDARD AND THE SMITHSONIAN

Robert Goddard shows off one of his A-series rockets in front of his workshop in Roswell, New Mexico, in the mid-1930s. He gave the Smithsonian a rocket almost identical to this one in 1935.

By the simple act of responding to a letter of inquiry from a pair of midwestern bicycle builders, Richard Rathbun had played a small role in launching the air age. In similar fashion, Charles Greeley Abbot helped get the Space Age under way. In late September 1916 he received a letter from Robert Hutchings Goddard, a professor of physics at Clark University. "For a number of years," the young academic began, "I have been at work upon a method of raising recording instruments to altitudes exceeding the limit of sounding balloons." Four long paragraphs later, he finally revealed that he had been investigating rocket propulsion.

A native of Worcester, Massachusetts, born in 1882, Goddard earned a B.Sc. from Worcester Polytechnic Institute (1908) and an M.Sc. (1911) and Ph.D. (1912) in physics from Clark University. After some important early work in electronics, the young professor began his work on rocketry and spaceflight. In 1914 he patented the design of both a multistage and a liquid propellant rocket and conducted an experiment demonstrating the ability of a rocket to function in space. The work was becoming ever more expensive, he explained to Abbot, and wondered if the Smithsonian could offer any support.

Abbot was immediately intrigued by Goddard's work. He had followed in Samuel Langley's footsteps, traveling to mountaintops and sending instrumented balloons aloft in an attempt to measure the solar constant, the total amount of solar energy reaching the Earth at the top of the atmosphere. Now he was hearing from a scientist who, in seven pages of exquisite detail, could explain precisely why a rocket was the ideal vehicle to loft instruments above the filtering atmosphere!

In less than a year, Abbot had arranged a $5,000 grant to support Goddard's first practical experiments in rocketry. No one was more pleased than the young scientist's mother. "I think that's the most wonderful thing I ever heard of," she remarked. "Think of it! You send the Government some typewritten sheets and some pictures, and they send you $1,000, and tell you they are going to send four more."

In the 1920s, Goddard sometimes placed the liquid-fuel rocket motor at the top of the vehicle. This picture was taken at Clark University in Worcester, Massachusetts, in 1929.

It was the beginning of a long and fruitful relationship. The Smithsonian published Goddard's classic treatise on rocketry, *A Method of Reaching Extreme Altitudes,* in 1919. The document was a serious engineering study filled with quadratic equations and tabular data designed to prove that existing solid-propellant rockets could carry instruments into space. The author did his best to understate the more sensational aspects of his study, confining his thoughts on the possibility of more efficient liquid-propellant rockets to a footnote and not even mentioning the possibility that human beings might one day ride on a rocket. The paper concluded, however, with a remark that it might even be possible to send a multistage rocket to the moon.

Goddard's valiant efforts to preserve his scientific dignity were a dismal failure. The shy professor and his "moon-going rocket" were front page news from the *New York Times* to the *San Francisco Chronicle* as a result of a Smithsonian press release about the publication. A Bronx promoter urged Goddard to consider the Starlight Amusement Park as a potential launch site, while a Hollywood agent cabled a request: "WOULD BE GRATEFUL FOR OPPORTUNITY TO SEND MESSAGE TO MOON FROM MARY PICKFORD ON YOUR TORPEDO ROCKET WHEN IT STARTS."

Retreating from the limelight, Goddard continued his work in relative secrecy and achieved a genuine milestone on March 16, 1926, when he sent the world's first liquid-propellant rocket sputtering aloft to a peak altitude of 41 feet above a Massachusetts cabbage patch. He continued to build, static-test, and launch his rockets for the next three years, gaining useful experience in the design and construction of pumps, valves, combustion chambers, nozzles, igniters, and the other bits and pieces that make up a rocket.

While the cost of Goddard's research quickly climbed beyond the Smithsonian's ability to support it, Charles Abbot introduced his younger friend to officials of the Carnegie Institution with deeper pockets. In the 1930s, when Charles Lindbergh persuaded Daniel Guggenheim and his son Harry to dig even deeper to fund Goddard's rocket work, Abbot remained as an adviser and facilitator. The Smithsonian published the scientist's second major work on rocketry, *Liquid-Propellant Rocket Development,* in 1936.

With the support of the Guggenheim Foundation, Robert Goddard was able to take long absences from his teaching duties and relocate to Roswell, New Mexico, where the clear skies and wide-open spaces provided the perfect conditions for test firing rockets. From 1930 to 1932 and again from 1934 to 1941, Goddard; his wife, Esther; and a handful of assistants lived and worked in relative isolation, developing the key systems required for a high-altitude rocket—improved combustion chambers, propellant pumps, cooling systems, and gyroscopic stabilizers. Goddard preferred to work alone, rejecting offers to cooperate with others—an approach that isolated him from other promising centers of rocket research in the United States, notably the talented group of graduate student experimenters exploring rocketry under the guidance of Theodore von Kármán at the California Institute of Technology.

In spite of his reluctance to share the details of his research or discuss the technical details of his rockets, Goddard did heed the advice of his strongest supporter, Charles Lindbergh, to preserve an example of his technology. As a result, the first artifact documenting the history of rocketry and spaceflight arrived on the Smithsonian's doorstep in 1935.

Goddard had begun the process with a letter to his old friend Charles Abbot offering to donate "one of the complete rockets that we have used in the stabilization development." He was offering one of the A-series rockets recently flown from his testing site near Roswell to the Smithsonian, "because of the help which it gave in the early stages of the work, when assistance was so important and at the same time so difficult to obtain." He stipulated that under no circumstances was the rocket to be shown to anyone without his permission, "or, in the event of my death, [that of] Mr. Harry F. Guggenheim and Colonel Charles A. Lindbergh."

Goddard's workers spent the next several weeks reassembling a typical A-series vehicle from parts of several surviving rockets, including A-5, which had been flown successfully on March 13, 1935. "The greatest

The smoke trail from this Roswell launch on August 26, 1937, shows Goddard's rocket's control system in action. He experimented with gyroscopes and moving rocket nozzles to solve the problem of stability and control.

height reached by these rockets," he explained to Abbot, "was somewhat over a mile, the greatest speed in flight over 700 miles per hour."

The rocket—carefully packed in a long wooden crate and looking much more sleek and streamlined than the sputtering craft that carried Buck Rogers and Dale Arden to the planet Mongo and back at the local Bijou every Saturday afternoon—arrived at the Smithsonian in November 1935. Measuring 15.35 inches long, 9 inches in diameter, and 21.5 inches across the fins, it was covered with a shiny, two-tone, aluminum-and-stainless-steel skin. One quadrant of the rocket and one side of several fins were painted bright red, so that any rotation could be observed as it rose into the air.

Having agreed not to display the rocket, Frank A. Taylor, curator of the Division of Engineering, reported on November 16 that the new acquisition had been moved into deep storage in a basement corridor of the Smithsonian Castle. There it would remain for over a decade. Goddard moved east again in 1941, and spent the war years in Annapolis, Maryland, developing rocket units designed to boost large seaplanes into the air. He died of cancer in August 1945.

Robert Goddard's historical stock began to rise with postwar interest in rockets and missiles. The rocket donated to the Smithsonian joined other examples of the professor's technology in an important postwar exhibition. In June 1960 the U.S. government awarded the Guggenheim Foundation and Esther Goddard one million dollars in consideration of the importance of the many patents issued to Goddard covering vital aspects of modern rocketry.

Visitors to the National Air and Space Museum have the opportunity to view a wide range of Goddard technology, from the world's oldest surviving liquid-propellant rocket to a Rube Goldberg device designed to indicate how much photographic flash powder would have to be exploded on the face of the moon to be visible from Earth. The A-series rocket received in 1935, a reminder of the role that the Smithsonian played in supporting one of the pioneers of the Space Age, has a place of honor near the entrance to the James S. McDonnell Space Hangar at the museum's Steven F. Udvar-Hazy Center near Dulles Airport. —*Tom D. Crouch*

Goddard and three key assistants examine one of his last and largest rockets in Roswell in 1940. This rocket, or one very similar to it, now stands in the Milestones of Flight gallery of the museum.

3 THE LONG ROAD TO A NEW MUSEUM

DOMINICK A. PISANO

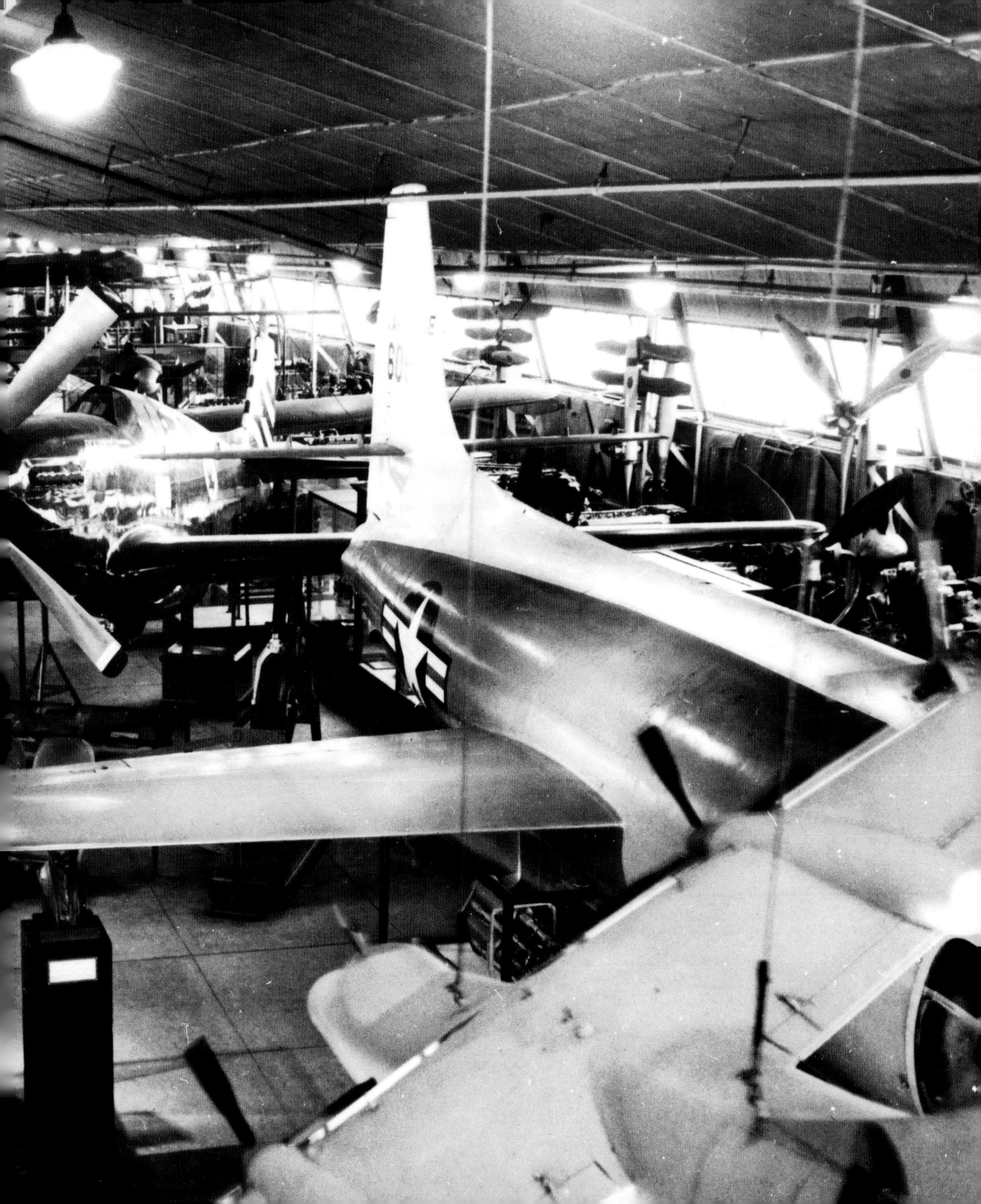

August 12, 1946, *was a momentous*

On that day President Harry S. Truman had signed Public Law 722, a bill sponsored by Jennings Randolph, Democratic congressman from West Virginia, creating the National Air Museum. Although there had previously been some interest in establishing a national museum for aeronautics, the legislation was the culmination of a quarter century of work by Garber, who, despite his relatively humble position, had been steadily collecting significant aeronautical treasures. / Garber had essentially kept the flame of a repository for aeronautical history alive at the Smithsonian Institution since his arrival in 1920, when he was given the task of exhibits *day in the life of Paul Edward Garber.*

preparator and charged with responsibility for repairing and maintaining museum exhibitions. By 1931 Garber had risen to the position of assistant curator in the Department of Arts and Industries in the Division of Engineering of the National Museum. During his years at the Smithsonian, Garber had made some astonishing acquisitions: the Curtiss NC-4, the aircraft that in May 1919 made the first flight across the Atlantic Ocean; the 1923 Fokker T-2, which made the first nonstop flight across the United States; the 1924 Douglas World Cruiser *Chicago;* Charles A. Lindbergh's 1927 Ryan NYP *Spirit of St. Louis;* and, in 1937, Wiley Post's Lockheed Vega *Winnie Mae.*

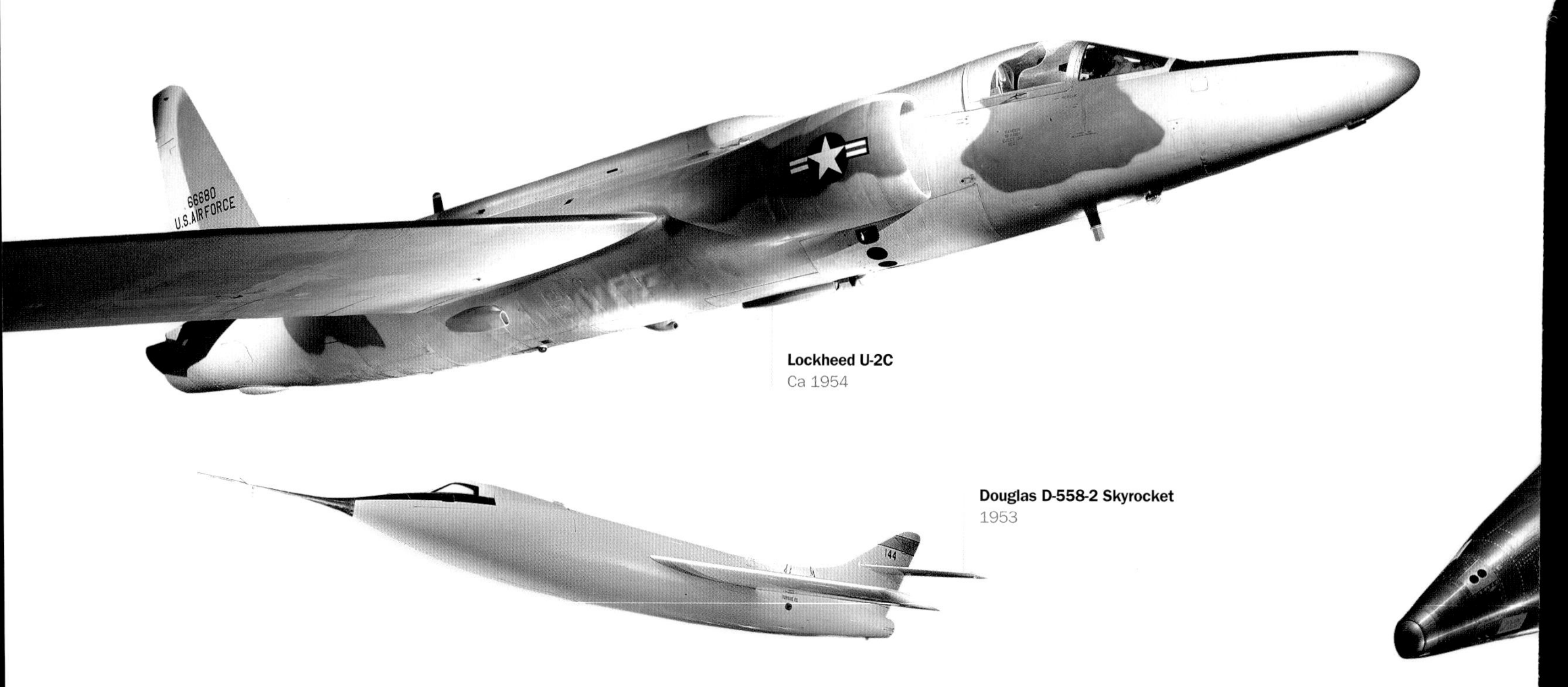

Lockheed U-2C
Ca 1954

Douglas D-558-2 Skyrocket
1953

FROM THE MUSEUM'S COLLECTION

1940s TO 1970s

Hiller Model 1031-A-1 Flying Platform
Ca 1955

Pitts Special S-1C
1948

North American X-15
1959

Grumman A-6E Intruder
1968

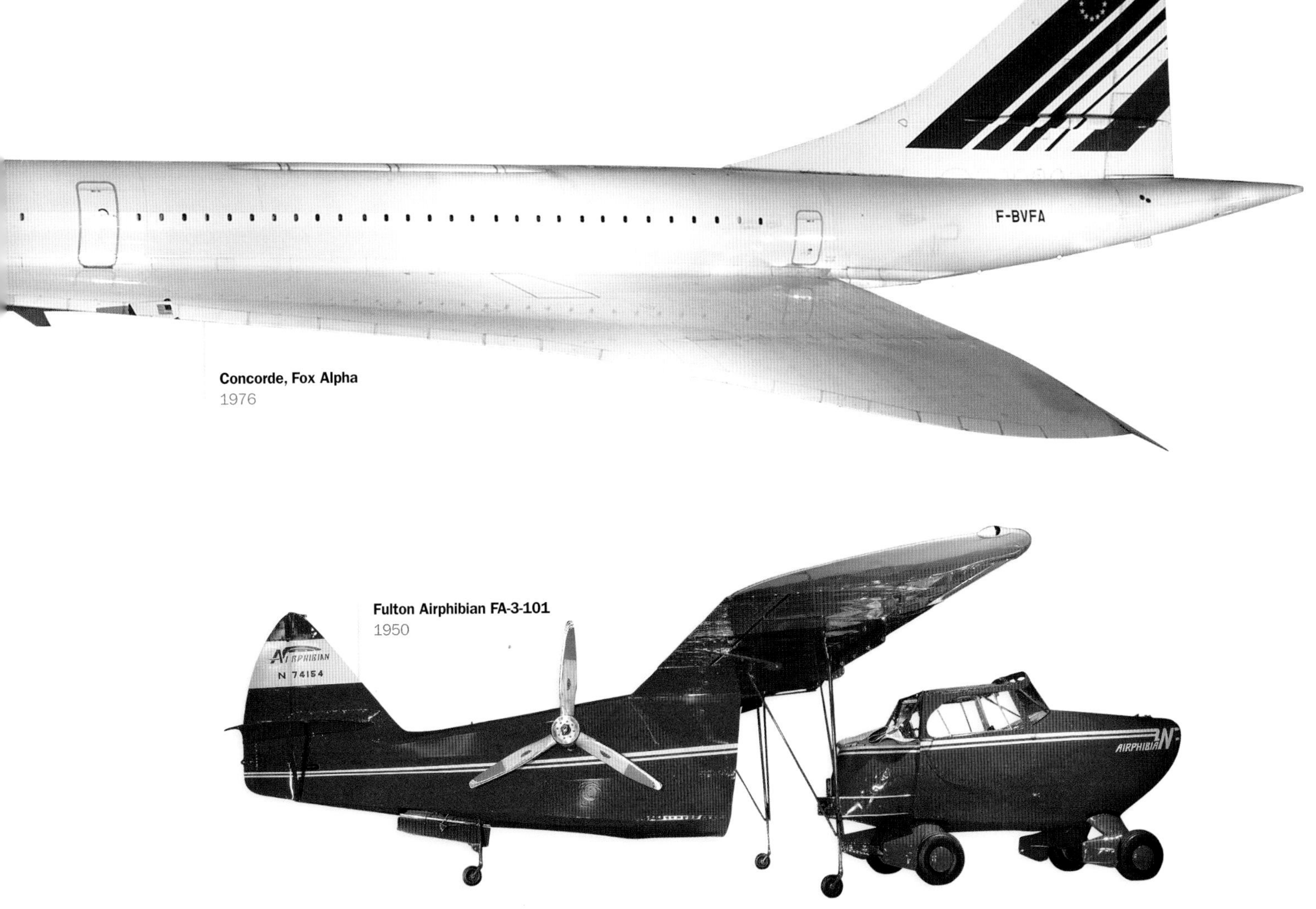

Concorde, Fox Alpha
1976

Fulton Airphibian FA-3-101
1950

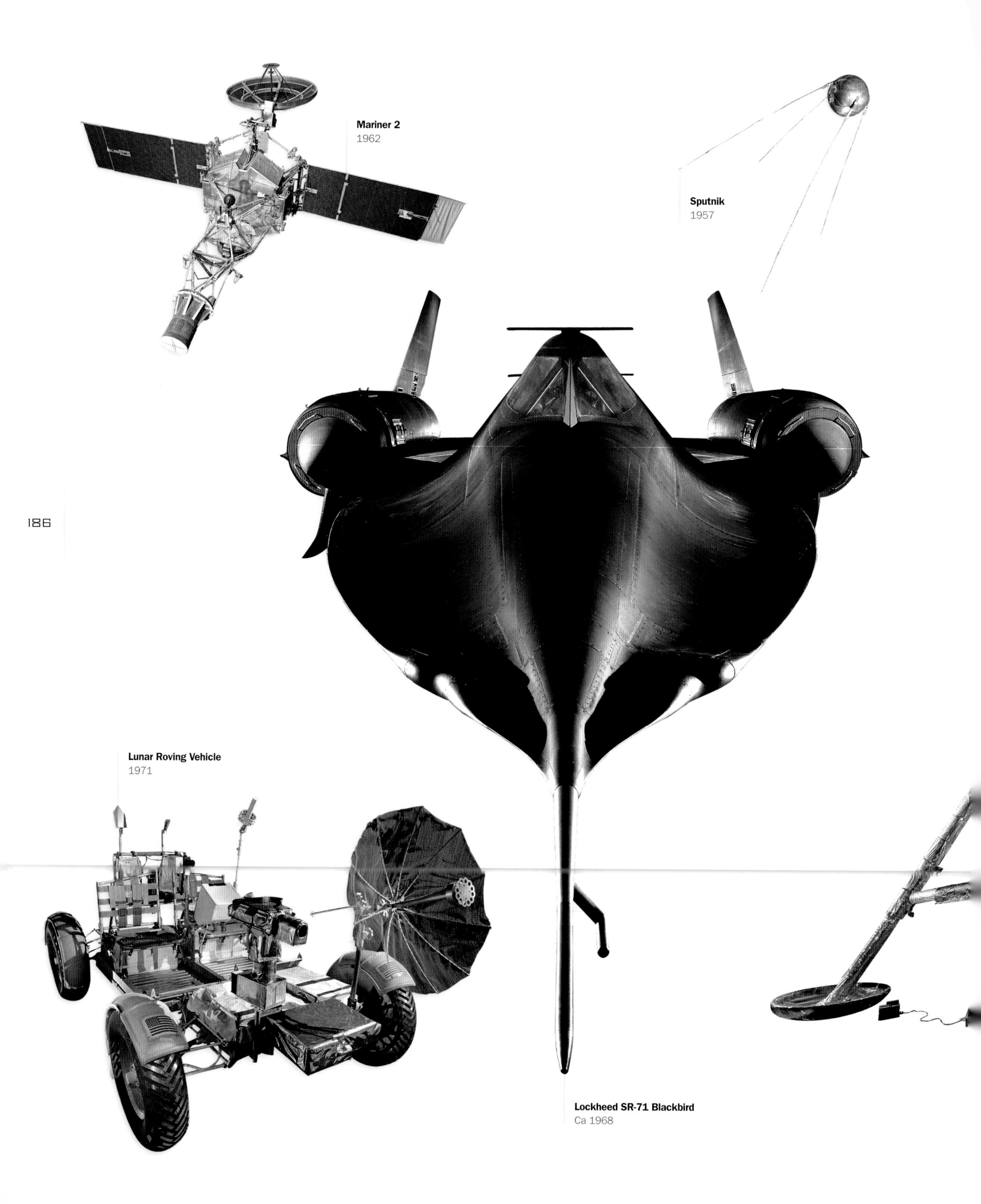
Mariner 2
1962
Sputnik
1957
Lunar Roving Vehicle
1971
Lockheed SR-71 Blackbird
Ca 1968

Apollo 11 Command Module
1969

Ranger
1964

Gemini 7
1965

Lunar Module #2
1968

Pioneer 10 / 11
1972

The immediate impetus for the museum had come from Gen. Henry H. "Hap" Arnold, wartime head of the U.S. Army Air Forces. In 1945 he began to collect representative surplus and captured World War II aircraft. He also gathered data that would support his case to obtain congressional authorization to construct a National Air Museum. Arnold's purpose was distinctly military: to prove to the public the importance of military aviation and help future generations become aeronautical engineers and military pilots. To get ideas for such a place, Arnold sent letters to the members of three of the largest aeronautical associations in the United States. In these letters Arnold asked the following questions: Where should such a museum be located; who should administer it; what the purpose of the museum should be. When Arnold's carefully researched and prepared materials reached the House of Representatives Committee on the Library in February 1946, the case for a national air museum was quite convincing.

Arnold had an ally in Randolph, an aviation advocate who had previously sponsored more than 20 aviation bills in Congress. Randolph introduced the museum bill before the Committee on the Library. Arnold testified that the U.S. Army Air Forces had made "great contributions during the war," that "young people were more interested in aviation than anything," and that Americans "must collect now" to retain a "full set of World War II aircraft." Although the original conception for the museum had been military and technological, under Randolph's sponsorship it became a place that would display aeronautical equipment of interest, but the museum's goal would be to "memorialize the national development of aviation rather than merely document the history of airplane technology." Randolph had wisely shifted the emphasis of the museum to a broader perspective, one that would deal with both military and civil aviation.

OPPOSITE, TOP: President Franklin D. Roosevelt sits with "Hap" Arnold, commanding general of U.S. Army Air Forces. Arnold ordered aircraft saved for display to the public: the basis for the National Air Museum.

OPPOSITE, BOTTOM: Paul Garber (right) accepts a ceremonial pen from Senator Jennings Randolph, who sponsored legislation creating the National Air Museum in 1946. Wayne W. Parrish, editor of *American Aviation,* looks on.

COAL BRIQUETTE Ca 1948 Of the type sent to West Berlin during the 1948-49 airlift to defeat the Soviet blockade

PAGES 180–81: The National Air Museum's Aircraft Building (the Tin Shed) shows crowded conditions around 1950.

The NAM advisory board had proposed various places for the National Air Museum, among them New York City, Dayton, Chicago, and Washington, D.C. But Alexander Wetmore, Secretary of the Smithsonian and vice chair of the National Advisory Committee for Aeronautics, argued strongly that the museum should be part of the Smithsonian complex and that it should be located in Washington, D.C. "After one hundred years of museum administration," he said, the Smithsonian "had acquired a certain 'know-how' in the operations of museums."

Wetmore also wrote to the chairman of the Committee on the Library and to the assistant director in charge of legislative reference at the Bureau of the Budget, emphasizing that the Smithsonian Institution was "air-minded." "I am pleased to state," Wetmore wrote, "that the Smithsonian Institution for sixty years had been definitely interested in matters concerned with the history of aviation, and that it has over a long period of time developed and maintained an important exhibition in this field in the U.S. National Museum in Washington." As part of the Smithsonian Institution, Wetmore argued, the National Air Museum could "develop logically" because the institution's administration was not subject to changes in political administrations. On July 9, 1946, the Committee on the Library submitted Report No. 2473, which stated "in the interest of simplicity the bill was amended to arrange to make a national air museum a bureau proper of the Smithsonian Institution. It provides that the Smithsonian Institution shall operate a national air museum with the advice of a board."

After the bill had passed the House of Representatives, it went to the Senate, where a discussion of it was held on July 31, 1946. Kentucky Senator Alben W. Barkley, who introduced the bill, made three significant points about it. First, he said, "the bill had been under consideration for a long time." Second, the House of Representatives had passed the bill, and it had strong encouragement from the Smithsonian Institution. Finally, Barkley pointed out that the bill had authorized the establishment of a board of advisers for the museum, but had not authorized the construction of a building. Three decades would pass, with numerous delays, political problems, institutional difficulties, and budgetary shortfalls, before the museum actually had a permanent structure in which to exhibit its collections.

THE WWII AIRCRAFT COLLECTION

A few years after the signing of the National Air Museum Act, Paul Garber became head curator of the museum, with responsibility for the preservation of the collection. One of the first tasks that Garber faced was the housing of Arnold's gift of World War II aircraft,

LIONS
L
INTERNATI

missiles, engines, and other objects. In November 1948, the National Air Museum received authorization to use a Douglas Aircraft Company plant at Park Ridge, Illinois, to store this collection temporarily, which consisted of 97 aircraft and 1,336 artifacts, including many rare specimens from Nazi Germany and imperial Japan. Park Ridge was the location of the Orchard Place Airport/Douglas Field, which had been constructed between 1942 and 1943 to manufacture Douglas C-54s during World War II. It is now Chicago's O'Hare Airport.

In 1951, as a result of the Korean War emergency, the United States Air Force served the museum with an eviction order. The collections stored at Park Ridge had to be moved by January 15, 1952, to make way for the reopening of the Douglas plant for manufacturing aircraft for the war, which had begun on June 25, 1950. Most of the collection went to outdoor storage and some to another building (T-7) in the Park Ridge complex. No sooner had Garber relocated the collection than he was served with yet another eviction order demanding that Building T-7 be evacuated. Garber and a small team of assistants literally worked around the clock to prepare the collection for transport to Washington. The Air Force general in charge of the Douglas plant supposedly threatened to move the aircraft "with a bulldozer" if Garber and his staff did not move quickly enough.

If the move out of Park Ridge was not difficult enough, Garber was faced with an even more serious problem: where to house the significant collection of military aircraft and artifacts that would make up a large part of the museum's collection | *to page 194* |

Military aircraft under tarpaulins are lined up outside a facility at Park Ridge, Illinois, a suburb of Chicago, for eventual delivery to the National Air Museum in Washington, D.C.

1702
237530
10

Aircraft stored outdoors in poor conditions at the Silver Hill facility included (left to right) the Northrop P-61, the Dornier Do 335, the Douglas XB-43, and the Douglas C-118. Although the museum tried its best to care for its artifacts, staff and funds were not available.

| *from page 190* | once it was shipped out of the Chicago area. This collection was much too large for the allotted display areas in the Smithsonian Institution complex, which consisted of the World War I–era Nissen hut behind the Castle called the Aircraft Building but nicknamed the Tin Shed, plus a space for exhibitions in the Arts and Industries Building.

Exercising a great deal of resourcefulness, Garber chartered a Piper J-3 Cub and surveyed the Washington, D.C., area for a suitable location. He eventually managed to obtain land in Suitland, Maryland, a couple of miles outside Washington in Maryland's Prince Georges County. Garber requested and received permission to use the Suitland land from the National Capital Planning Commission and the General Services Administration. At the site, known as Silver Hill, he saw to it that six prefabricated Butler buildings were erected to house the Park Ridge collections.

Garber started with merely the land. He had to "scrounge," he said, which meant he had to beg, borrow, or steal to construct the buildings to house the collection. With no hope of obtaining financial help from the Smithsonian Institution, Garber, in his inimitable style, cajoled the Army, Navy, Air Force, the local fire department, and a local concrete company to assist him. From the Army at Fort Belvoir, Garber recalled, "I got to another engineering officer, told him of my need for a big bulldozer, he sent one to our Facility, its blade was raised against a big tree, the bulldozer pushed, the tree fell away, its roots came up and a working party started sawing. In due time, all trees marked for removal were in a huge pile, [and] with the help of a few tires a fire was started, [and] with the Silver Hill Fire Department next door, every precaution was taken, the trees became smoke and cinders, and we had space for buildings and roads. We named one Randolph Road for Jennings and one Arnold Avenue for Hap."

OPPOSITE, TOP: The rear fuselage of the Arado Ar 234B Blitz (Lightning), the world's first operational jet bomber and reconnaissance aircraft, at Park Ridge, Illinois. Captured by the Allies, it became part of the Air Force collection given to the Smithsonian.

OPPOSITE, BOTTOM: Tail section of the Douglas XB-42A Mixmaster at Park Ridge in 1951. A rear-engined experimental bomber designed for high speed, it made its first test flight on May 6, 1944. When jet aircraft were developed, the Army abandoned the project.

AVIATOR'S HELMET Ca 1948 Type H-1 U.S. Navy, the first standard hard helmet issued to Navy jet pilots

Before long the National Air Museum complex at Suitland consisted of 16 metal Butler buildings. Garber and his small crew had to move the collection inside the buildings. Garber later recalled that in "our strenuous efforts to save those aircraft from the ravages of the weather and vandals, I credit Walter Male for his excellent knowledge of aircraft structures, mechanical methods, and tools, and ability to work with other mechanics, and even to enlist the help of our guards in getting the aircraft at Park Ridge disassembled, boxed, braced, and moved, then coming with them to Silver Hill to direct their restoration and assembly, and placing them on exhibit insofar as our very limited space afforded."

At about the same time that the Air Force was setting aside aircraft and artifacts for the National Air Museum collection, the U.S. Navy established a museum storage area at the Naval Air Station in Norfolk, Virginia.

During its early years, the National Air Museum received, in addition to the collections it had obtained from the Air Force and the Navy, two historically significant aircraft and a collection of archival documents, books, and other materials on the history of aeronautics that formed the basis of its nonaircraft collections. In 1949 the museum acquired the Boeing B-29 *Enola Gay,* the aircraft that dropped the first atomic bomb in August 1945, as part of the Air Force collection. After the war, the *Enola Gay,* which had taken part in the Bikini Atoll tests in 1946, was flown to Davis-Monthan Army Air Field in Arizona to be stored. Later the Air Force flew the *Enola Gay* to Park Ridge, Illinois, and transferred ownership to the Smithsonian on July 4, 1949. Because the National Air Museum had no place to display or store the aircraft, it was moved to Pyote Air Force Base, Texas, between January 1952 and December 1953, and then to Andrews Air Force Base, Maryland. It stayed at Andrews in outdoor storage until 1960–61. Because it had been outside, the aircraft had been damaged by weather and vandals, and the NAM decided to disassemble it (a year-long process) and then moved it indoors at Silver Hill.

The Bell X-1 *Glamorous Glennis,* the aircraft in which Capt. Charles E. "Chuck" Yeager had broken the sound barrier on October 14, 1947, made its last flight on May 12, 1950, for an appearance in the RKO film *Jet Pilot.* On July 10, 1950, the USAF's Office of the Chief of Staff ordered the X-1 to be prepared for delivery to Boston by August 23, 1950, to Boston's Logan International Airport. On August 26 and 27, 1950, it was placed on display at the Air Force Association's fourth national convention and National Air Fair in Boston. On September 4, the aircraft was flown to Andrews Air Force Base, strapped to its Boeing B-29 mother ship. The next day it was detached from the B-29. With the help of Bolling Air Force Base personnel, the aircraft was brought on a flatbed truck through the streets of Washington. Paul Garber was on hand when the X-1 was

T2-1010
350224
8433

installed at the National Air Museum's Aircraft Building on September 18. With help from the Bell Aircraft Corporation, the X-1 was displayed mounted on a pedestal.

Because the X-1 was displayed in close quarters, with other aircraft surrounding it, the public did not have an unobstructed view of it. In 1959, the NAM decided to move the aircraft to a more suitable space and to create an exhibition for it. Plans included pouring a concrete base that was 4 feet by 4 feet by 2 feet, which would hold a steel plate that was 15 feet by 15 feet by 6 inches into which would be placed a 5-inch steel tube mounting. The aircraft was placed on the mounting at an angle of approximately 45 degrees, surrounded by a stainless-steel railing and the appropriate labeling. An XLR-11 (6000 C-4) rocket engine, the type that powered the aircraft during its test-flight period, was displayed along with the aircraft so that the public would have an idea of what the X-1's propulsion system looked like outside the aircraft.

In the early 1960s, through the efforts of Paul Garber, the museum was able to acquire the collections of the Institute of the Aeronautical Sciences, one of the two predecessors of the American Institute of Aeronautics and Astronautics. Aware that the IAS was running out of room to house its various collections at its headquarters in New York City, Garber in December 1954 presented a convincing proposal on how the collection could be transferred to the National Air Museum. Finally, in July 1963, Garber himself transported numerous loads to Washington. In his notes was an inventory of the IAS collections, which included 400 cartons of books; 160 pamphlets, brochures, and biographical items; 10 cartons of the Sherman Fairchild photographic collection; 39 cartons of miscellaneous photographs; 74 cartons of motion picture film; five boxes of material from the Bella C. Landauer Collection of Aeronautical Sheet Music; 18 cartons of model aircraft; 29 boxes of commemorative medals; 171 early aeronautical prints from the Harry F. Guggenheim Collection; 2 packages of posters; and 4 watches that belonged to the Wright brothers. Today, many of these materials—notably the Landauer, Fairchild, and Guggenheim collections—are considered to be among the gems of the National Air and Space Museum's holdings.

LUNCH BOX
1960
In the late 1960s and 1970s, the museum staff's lecture series opened with the symbolic placement of this lunch box on the table.

OPPOSITE, TOP: An interior view of the Air Force storage facility at Park Ridge. In the foreground is the Heinkel He 219 A-2/R4 Uhu *(Eagle Owl)*, a night fighter-reconnaissance aircraft brought to Freeman Field, Indiana, for flight testing and then transferred to the National Air Museum.

OPPOSITE, BOTTOM: Another view of Park Ridge. At center is a captured V-2 missile, the world's first ballistic missile. The V-2 (Vergeltungswaffe Zwei, or Vengeance Weapon 2) was used in German attacks on various cities in Europe, including more than 1,300 on London.

SEARCH FOR A PERMANENT SITE

Having accomplished the difficult task of moving the large military aircraft and artifact collections from Park Ridge to Suitland, Maryland, the National Air Museum's next task, as recommended by the museum's advisory board, was to find a suitable place to construct a museum building that could house all of the aircraft collections. The Smithsonian completed basic studies of the museum's building requirements with the assistance of the General Services Administration's Public Building Service, and the results had been submitted to Congress for approval in March 1950. The initial plans for the museum that dated back to 1947 called for 500,000 square feet of floor space to exhibit 200 aircraft, with room for future expansion. The NAM advisory board had selected Daingerfield Island, near Washington National Airport, as the ideal site for the museum. In view of the projected size of the museum, substantial parcels of land were required.

Other suggested sites for the National Air Museum were Arlington Farms, adjacent to both Arlington National Cemetery and the abandoned Washington-Hoover Airport, and Theodore Roosevelt Island. These large tracts of land were enthusiastically sought by the advisory board in the years immediately after World War II, but they were designated for other projects. Among other sites of interest were Beltsville, Maryland, and Andrews Air Force Base, neither of which were close to the National Mall, where the majority of the Smithsonian buildings were located.

In June 1953 the advisory board reviewed the site proposals and came to the conclusion that "a permanent building for exhibition and administration of the National Air Museum's outstanding collection of aircraft" should be "located in the vicinity of the present group of Smithsonian buildings." The advisory board's rationale for the decision was that only a building near other museums would "be of maximum service to the millions of visitors who will want to see this museum." In addition, the advisory board believed that the choicest artifacts could be put on display in the main building and a separate facility would be reserved for the majority of the collection, which would be available for research and study.

In 1955, the National Capital Planning Commission agreed to a site for the museum across the street from the Castle on Independence Avenue, S.W. (now the site of the Energy Department's Forrestal Building). The Aircraft Industries Association and the Air Transport Association | *to page 202* |

Three U.S. military rockets stand on display before the Smithsonian Institution's Arts and Industries Building, early 1960s: (left to right) the U.S. Army Jupiter-C satellite launcher, the U.S. Navy Vanguard satellite launcher, and the U.S. Navy Polaris A-3 missile. In those days, the area outside the museum building became known as Rocket Row.

JUPITER C
U S
ARMY

Lou Purnell was a member of the historic 99th Fighter Squadron, the first all–African-American combat unit in the U.S. Army Air Corps (later the U.S. Army Air Forces) during World War II. In 1967 he joined the curatorial staff of the National Air and Space Museum. His specialties were spacesuits and space food. Here are some of his recollections, given in an interview for the Smithsonian Archives, of the Apollo 11 moon landing on July 20, 1969.

LOUIS R. PURNELL

ON APOLLO 11

On the first floor, right in front of Arts and Industries in that foyer . . . —what do they call it now? The atrium. That's where everything was set up. The news reporters, those that would come on with live broadcasts from television, radio, newspaper reporters, anything that you can think of for such an event, those people were there. People from all the media were there all night long.

Not only did I live in high hopes [of getting, and not only was I] almost assured of getting that spacesuit that was on the moon at the time, but [I also hoped that I would have a chance] of meeting the people.

The night that the astronauts landed on the moon, I've never seen before a gathering of important people. There was Roger Mudd . . . Walter Cronkite . . . Dan Rather.

In fact, Dan Rather was in the next office to mine, and I learned a lot about him.

He's a diligent worker. The other reporters were there, talking among themselves. Dan Rather had his sleeves rolled up, and he was setting that typewriter on fire in Fred Durant's office. [Durant was the museum's assistant director for astronautics.]

I went in and told [Rather], "When are you going to let up? Aren't you going to eat lunch?"

He handed me some money and said, "Go out to McDonald's and get some hamburgers or chicken or something and give it to the rest of the staff downstairs and bring me something."

But just to meet those people that night. We stayed up all night. I took a cot down to my office and a large coffee pot and served a lot of the staff coffee. We stayed overnight just for that moment.

As I said before, just high hopes of getting all that equipment that I saw being used on the moon made my job more exciting, and when I actually did touch the Apollo 11, I could say that I had seen it launched.

We'll never get a lunar lander because they are left on the moon. [The museum did get the unflown lunar module (LM-2) in 1971, reconstructed to look like Armstrong and Aldrin's LM-5 *Eagle,* which was left on the moon.]

But the spacesuits and the people who participated, all the way from the first who touched the moon to the last Apollo landing—just meeting those people, very exciting.

| from page 197 | donated $25,000 for an architectural study by the prestigious firm of McKim, Mead and White. The architects' plan was expansive, calling for a building that would extend the entire block along Independence from Ninth to Tenth Streets S.W., and another block south to C Street. Inside the building, the Wright Flyer would be given a place of prominence and enshrinement. The plan also called for many of the military aircraft in the collection to be displayed, including the *Enola Gay.* Other galleries were devoted to a hall of fame and an industry exhibit.

But the Independence Avenue site got caught up in the politics of redevelopment. In 1946, Congress had passed the District of Columbia Redevelopment Act, creating the Redevelopment Land Agency (RLA) to redevelop parts of the city. In 1951, the RLA began to devote its attention to the redevelopment of the city's southwest quadrant, an impoverished black neighborhood. What evolved was a plan to raze all the existing structures in the area picked by the museum's advisory board, to build housing for federal workers and a complex of offices and businesses that would be known as L'Enfant Plaza. In view of the RLA's plan for the area, the NCPC eventually changed its mind about allowing the site to be used for the National Air Museum, and instead approved it for the L'Enfant Plaza complex.

THE HOPKINS AND THE JOHNSTON YEARS

In February 1958, the National Air Museum got its first official director, Philip S. Hopkins. (Before Hopkins, the museum had been under the direction of Carl W. Mitman, who from 1948 until his retirement in 1952 had been assistant to the Secretary for the National Air Museum. In 1952, Paul Garber was given the title head curator, which made him in effect head of the NAM.) Hopkins had been a patent attorney in Binghamton, New York, and one of his first clients was Edward Link, the inventor of the Link Trainer, the first elementary aircraft simulator. Hopkins eventually became Link's attorney, and later a vice president of Link Aviation. Hopkins was intensely interested in aviation education. In 1950 he had founded the National Aerospace Education Council. Before becoming National Air Museum director, Hopkins was a professor at Norwich University in Northfield, Vermont, where he had established a department of aviation.

In September 1958, Senator Clinton P. Anderson, aircraft manufacturer Grover Loening, and Smithsonian Secretary Leonard Carmichael helped push a bill through Congress that set aside a plot of land between Independence Avenue S.W. and Jefferson Drive, and between Fourth Street and Seventh Street S.W., across from the National Gallery of Art. This plan had received the blessing of the NCPC, the Bureau of the Budget, the General Services Administration, and the Commission of Fine Arts, all of which had a great deal of influence over the construction of public buildings in the District of Columbia.

Nevertheless, a heated battle took place in Congress over the National Mall site for the museum because advocates of the proposed National Capital Center for the Arts (now the John F. Kennedy Center for the Performing Arts) believed that the NAM would be too militaristic and that an arts center should be constructed on the site. Leading the arts center advocates was Senator J. William Fulbright. Senator Anderson's bill defended the NAM site's right to be on the mall, arguing that "aviation is one of the few great human achievements that have originated, developed, and come to possibly their greatest flowering in the United States. Some of the greatest and most imaginative minds of American science have worked on the problems of aviation."

Also in 1958, President Dwight D. Eisenhower signed a bill authorizing the Smithsonian Institution Regents to propose plans for the National Air Museum's building. Having secured the National Mall site, Hopkins turned his attention to modifying the 1955 plan for NAM, obtaining congressional funding for construction, and selecting an architectural firm to design the building.

In 1960 preliminary floor plans and interior concepts were approved, and in 1963 Congress allocated $511,000 in planning funds. The St. Louis architectural firm of Hellmuth, Obata + Kassabaum, now called HOK, was selected to make the final design. The so-called Space Race had begun, and the building requirements specified that it "was to provide exhibit areas with a degree of flexibility which would permit the display of objects ranging from subminiaturized instruments up to very large complete aircraft, space vehicles, and rocket boosters." Also required was "a building whose external appearance would be compatible with its location on the mall in the proximity of the Capitol and the National Gallery of Art." Unfortunately, Hopkins left the

OPPOSITE, TOP: An Apollo program review in September 1962 brings together a historic group: (left to right) President John F. Kennedy; NASA Marshall Center Director Wernher von Braun; NASA Administrator James Webb; Vice President Lyndon Johnson; Secretary of Defense Robert McNamara; Kennedy's science adviser Jerome Wiesner; and Director of Defense Research and Engineering Harold Brown.

OPPOSITE, BOTTOM: An aerial view shows the museum's Silver Hill, Maryland, storage site, circa 1960. It was renamed the Paul E. Garber Facility in 1980, in honor of the man who created it.

PILOT'S KNEE BOARD 1960 Used by Scott Crossfield to take notes during the 12th flight of X-15 rocket aircraft, December 6, 1960

directorship in 1964 without anything but preliminary plans for a museum building.

In September 1964, Carmichael named S. Paul Johnston director of the National Air Museum. Johnston had been trained as an aviator in World War I, graduated from MIT with a degree in mechanical engineering in 1921, and had served with the Naval Air Transport Service during the Second World War. From 1930 to 1940 he had been editor in chief of *Aviation* magazine. From 1946 until he accepted the NAM directorship, he had led the Institute of the Aeronautical Sciences.

During his tenure, Johnston recruited a panel of advisers to prepare a report, which appeared in January 1965, titled "Proposed Objectives and Plans for the National Air and Space Museum" (one of the first references to the expanded name Congress formally enacted into law in 1966). The report, as Johnston saw it, would be to "study overall objectives and plans for the National Air and Space Museum consistent with its legal mandates, its long term objectives, its potential audiences and its educational opportunities, optimized for the period 1969–1970 when (hopefully) the doors may be expected to open." The advisers Johnston chose for this project were: Preston R. Bassett, a past president of the IAS; Frederick C. Durant III, past president of the American Rocket Society (the other organization that merged into the AIAA in 1962–63) and of the International Astronautical Federation; William Littlewood, past president of the IAS and of the Society of Automotive Engineers; and Addison Rothrock, a professional engineer with longtime experience at the National Advisory Committee for Aeronautics.

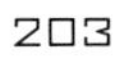

Johnston's conception of the museum was decidedly in favor of space science and technology as an integral part of the museum's program—a reflection of the country's increased interest in space exploration and the race with the Soviet Union that had begun with the launch of Sputnik on October 4, 1957. The report even included an organization chart that listed an associate director for astronautics, with responsibility for space and reentry vehicles, launch and propulsion systems, control, communications and navigation systems, and life support and related systems. To this end, Johnston selected Frederick C. Durant, one of the advisers who had helped draft "Proposed Objectives and Plans," to be the first assistant director for astronautics.

Durant had been born to an old Philadelphia family, was educated at Lehigh University as a chemical engineer, and had served as a naval flight instructor and test pilot during World War II. After the war he had worked at Bell Aircraft and afterward became director of engineering at the U.S. Naval Air Rocket Test Station (NARTS), in Dover, New Jersey. After NARTS he officially became a private consultant, even while continuing to serve in the Naval Reserve as an officer; he was actually working as a Central Intelligence Agency analyst specializing in tracking foreign rocket activity and personnel. At the end of the 1950s, Durant was possibly the most well-connected space authority in the world. He was a close friend of the German-American rocket engineer and space visionary Wernher von Braun. He had worked with von Braun and the Office of Naval Research on Project Orbiter, the first practical proposal to put a U.S. satellite into orbit. Project Orbiter lost out in 1955 to Project Vanguard.

OPPOSITE, TOP: Paul Garber (in cockpit) oversees the installation of the Kugisho MXY7 Model 22 Ohka (Cherry Blossom) kamikaze suicide bomb in the Arts and Industries Building. It was nicknamed by Americans *Baka* (Japanese for "idiot" or "fool").

OPPOSITE, BOTTOM: Garber (left) guides the installation of the Bell X-1 rocket plane into the Aircraft Building in 1950. The X-1 was one of the most significant acquisitions of the museum in the first years of its existence.

RUSSIAN NESTING DOLLS 1962 Francis Gary Powers, the U-2 pilot shot down by the U.S.S.R. in 1960, purchased these dolls after his release from imprisonment.

Johnston was quite optimistic that a building for the museum could be constructed and that the new museum would be a place where "every visitor who goes through its halls, whether his stay be brief or protracted, should leave the museum more knowledgeable than when he entered. Young people in particular should be encouraged to learn more about the subject matter on view and be stimulated toward eventual participation in aerospace science and technology. Hopefully, every visitor should be consciously or unconsciously conditioned to better relate his own life and work to the rapidly moving age in which he lives."

Johnston's five-year plan for the museum was thoughtful, forward thinking, and geared toward developing specialists within the museum who were "highly competent in aerospace science and technology so that they can communicate competently with technical and scientific personnel in industry." He also developed a number of criteria for the display of objects: evidence of technical importance, historical importance, biographical importance, public impact, practical utility, importance in the chronological pattern, general (or sentimental) interest, as well as availability.

On July 19, 1966, President Lyndon B. Johnson signed the National Air Museum Amendments Act, which changed the museum's name to the National Air and Space Museum. Five departments in the museum were set up: aeronautics, astronautics, education and information, exhibits, and administration. In 1967, the museum concluded an agreement with the National Aeronautics and Space Administration to preserve and display spacecraft and related materials from the national space program. The following year Johnston and NASA administrator James Webb signed an agreement that specified that NASA would agree to turn over to the museum "such artifacts currently under NASA control and which become available in the future, after technical utility to NASA or other government agencies has been exhausted and post flight examination has been effected."

Even though the NASA–NASM Agreement formalized the acquisition of space artifacts, the NAM had been collecting significant objects from the U.S. space program since the late 1950s, and the Smithsonian Institution had shown an early interest in the work of Robert Goddard, the American rocket pioneer. Paul Garber had assisted the Guggenheims in creating an exhibition on Goddard during the late 1940s, and he collected all of the objects in the exhibition when it closed. Later he brought the first spacecraft into the NAM collection. The Vanguard launch vehicle and satellite arrived in 1958, followed in 1959 by a Jupiter-C rocket like the one that orbited the first American satellite, Explorer I. (The acquisition of military missiles resulted in an outdoor display nicknamed Rocket Row, which was located on the mall side of the Arts and Industries Building.) In 1961, the National Air and Space Museum received its first human spacecraft, Alan Shepard's Mercury capsule *Freedom 7*, followed 18 months later by John Glenn's *Friendship 7*.

The National Air Museum Amendments Act was also significant because it authorized the construction of a new museum building that would be 784 feet | *to page 210* |

JAPANESE BAKA-BOMB

BELL Aircraft
GLAMOROUS
GLENNIS

The X-1 under construction at Bell Aircraft Corporation in Buffalo, New York

On October 14, 1947, flying the Bell XS-1 (later designated the X-1) rocket plane, Capt. Charles "Chuck" Yeager, USAF, became the first pilot to fly faster than the speed of sound. He reached Mach 1.06, 700 miles an hour, at an altitude of 43,000 feet over the Mojave Desert near Muroc Dry Lake, California. Yeager's flight proved that an aircraft could exceed the speed of sound (about 660 miles an hour at 40,000 feet) and shattered the myth of an invincible "sound barrier."

BELL X-1

FIRST TO BREAK THE SOUND BARRIER

Because there was insufficient data on flight at or near that velocity, designers at Bell Aircraft Corporation decided that a supersonic aircraft should be shaped like a .50-caliber bullet. Ballistics tests indicated that high-powered bullets traveled routinely at speeds in excess of the speed of sound. That is why the X-1 has a bullet-shaped fuselage and a pointed nose.

The X-1 was dropped at about 20,000 feet from the bomb bay of a Boeing B-29. The aircraft then used its rocket engine to climb to its test altitude and accelerate.

Capt. Charles E. "Chuck" Yeager poses in the cockpit of *Glamorous Glennis* at Muroc Air Force Base, California, in May 1948. OPPOSITE: The instrument panel inside the cockpit of the X-1

After the Smithsonian officially accepted the X-1 from the Air Force in August 1950, it was placed in the Aircraft Building the next month. In 1975 it was moved to the new museum and placed in the Milestones of Flight gallery.

On his visits to the museum, Yeager frequently complained that the aircraft had the wrong paint scheme. It had come as it had appeared in its final flight for the film *Jet Pilot*. Consequently, staff painter Danny Fletcher, working from a scissors-lift crane and using an ordinary roller, restored the airplane's original all-orange paint scheme. To complete the job, the Air Force star and stripes insignia—originally on the upper-left and lower-right wing surfaces but then removed—was restored.

The X-1 now looks just like it did the day it broke the sound barrier.

| from page 204 | long, 250 feet wide, and 97 feet tall. However, a clause was attached to the Smithsonian's budget recommending that the appropriation for the building be withheld until spending on the war in Vietnam was reduced. Although the clause did not have the force of law, it was treated by Congress as though it did. In 1968, the Smithsonian attempted to bypass the funding block by requesting permission to build a parking garage underground, collect fees, and build the main part of the museum later. Congress refused this request in 1968 and again in 1969. Johnston's initial optimism about his being able to make the National Air and Space Museum building a reality became clouded with pessimism.

In September 1969 S. Paul Johnston retired as director of the museum, but on his way out he gave a scathing speech to the Aero Club of Washington expressing his disillusionment with Smithsonian leadership and criticizing its inaction in supporting the museum's request for a building. He charged that Smithsonian officials had neglected his museum "in favor of projects less than vital to the public interest" and that they favored "scholarship and art over 'more practical, hardware-oriented technologies of flight.' "

He had begun his tenure "bright-eyed and bushy-tailed over the prospect of creating here in Washington a great new facility on behalf of the government and the aerospace industry," he said, and:

Now, five years later, I make my valedictory address in an atmosphere of frustration and personal disappointment. The bright outlook of 1965 has receded further upstream each year. At the moment, the prospects for a new facility are at least as far in the future as they were then. We can't yet see around that corner. Disappointing as all this may be, we are all aware of the underlying reasons. This country's dollar commitments at home and abroad for urgent domestic and military programs are currently the overriding considerations. This is not exactly the time to go to the Congress for some $50 million for a museum building in Washington. In fact, during the hearings on our enabling legislation [HR 6125 passed and signed by President Johnson in July of 1966], the Smithsonian was enjoined from coming to the congress for construction funding until after the settlement of the Vietnamese War.

Johnston referred pointedly to what he called "other problems":

Most of us who have been in this aviation business for many years develop a certain amount of self delusion. In our enthusiasm for our own activities, we think that everybody *is impressed by—and vitally interested in—our mechanical birds and their accomplishments. It is sometimes a shock to find that this "just ain't necessarily so." Especially around a place like the Smithsonian, there are any number of "ologies" and sociologically-oriented disciplines whose practitioners consider aircraft only as a means of getting out to the remote boondocks to study baboon behavior, or to look into the private life of the green spotted frog of the upper Amazon—and* spacecraft *are important only as vehicles to carry biological or astrophysical experiments. Now these things are all well worth doing—but it comes down to a question of* priorities*—and when (as at present)* money *and* manpower *are being rationed, the question is—where do money and man-power go first? Answer:—Not to the Air and Space Museum. Unfortunately, from our point of view, the current art and "ology"-oriented management of the Smithsonian appears to favor sculpture gardens, folk art (both performing and static), and elaborate housing for the scholarly, over the more practical, hardware-oriented technologies of flight.*

One of Johnston's chief complaints was that the National Air and Space Museum was placed organizationally under the Smithsonian Institution's assistant secretary for History and Art. "The incumbent," Johnston stated, was "a PhD—specialist in the political history of England in the 18th Century. He takes some pride in the fact that he has never come within miles of the Pentagon—physically or spiritually. He has little personal interest in aerospace matters, and yet he is representing us to the Upper Councils of the Smithsonian on our programs and priorities. I protested this arrangement when it was first announced about a year ago on the grounds that we were substantially more akin to science and technology than to history and art but to no avail."

Johnston noted that several million dollars were spent by the Smithsonian administration "simply to provide a properly elegant atmosphere for Visiting Scholars" (a probable reference to the new Woodrow Wilson Center), in contrast to the mere $600,000 provided to the National Air and Space Museum out of a total Smithsonian budget of $30 million. He warned that "there is presently a real danger that, unless you | to page 214 |

OPPOSITE: A 1958 television interview in which Frederick C. Durant III (right) helps demonstrate rocketry. Durant became the museum's first assistant director for astronautics in 1964. The man at left is unidentified.

ADVERTISING BROCHURE Ca 1950 The Beechcraft Bonanza B35 with its "V" tail became an iconic private aircraft of the post–World War II era.

DANGER!
ROCKET FUEL

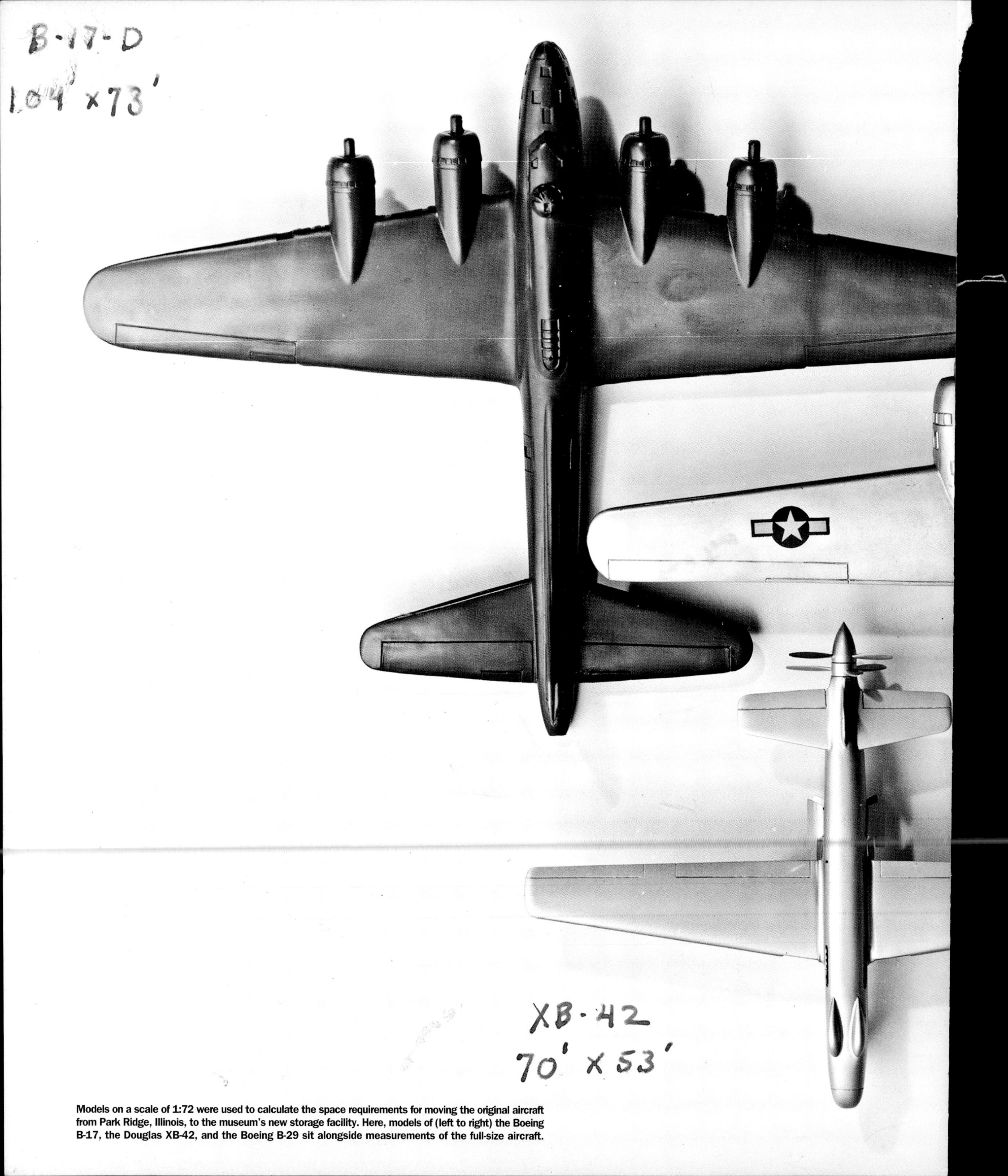

Models on a scale of 1:72 were used to calculate the space requirements for moving the original aircraft from Park Ridge, Illinois, to the museum's new storage facility. Here, models of (left to right) the Boeing B-17, the Douglas XB-42, and the Boeing B-29 sit alongside measurements of the full-size aircraft.

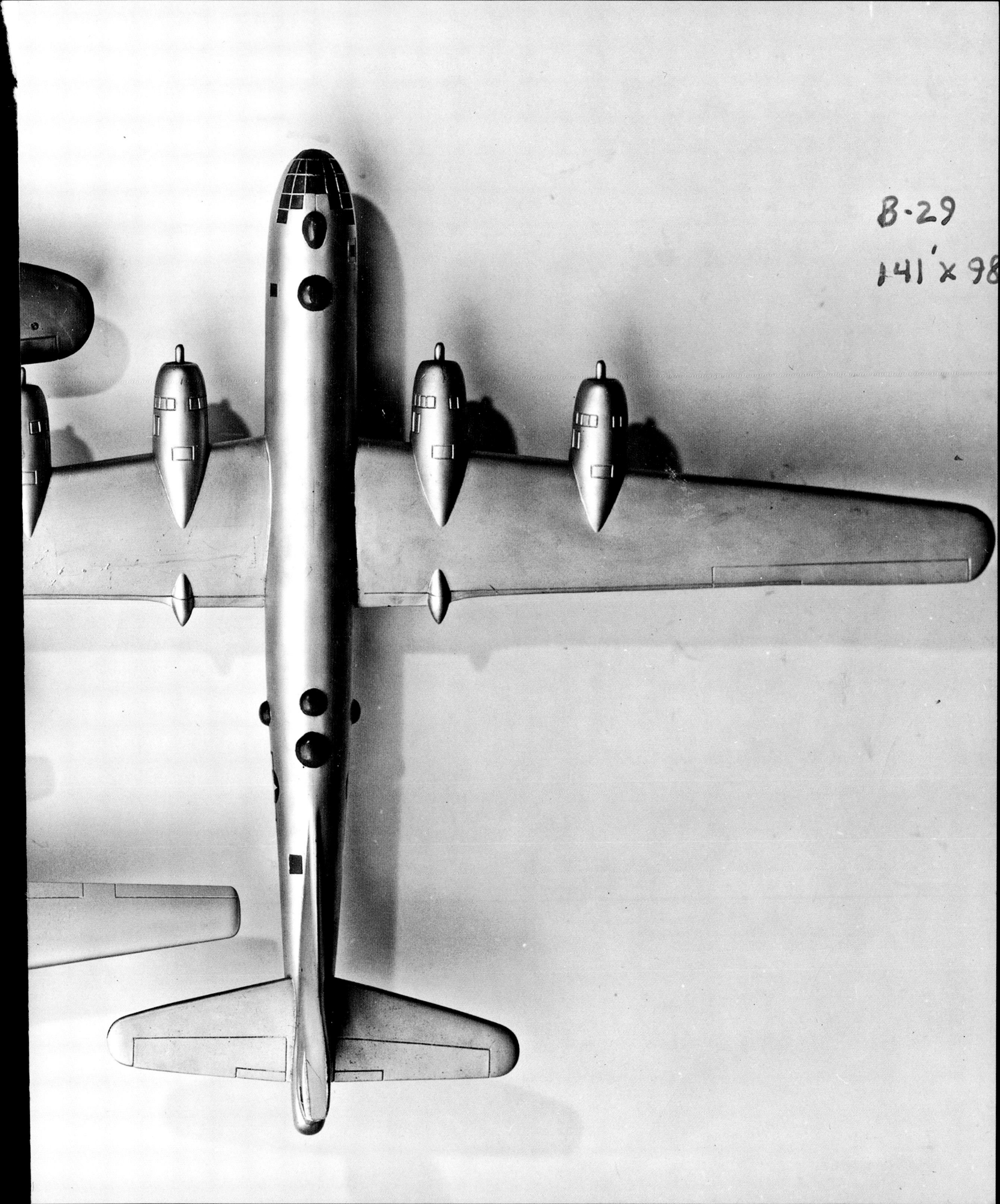
B-29
141' x 98

| from page 210 | people in the industry and in scientific and technical societies take a *real* interest in its future and *take positive action to insure* that future, it [the National Air and Space Museum] may never get off the ground."

Johnston ended his speech with a plea for help to the aerospace community. "I am not about to suggest *how* you go about it," Johnston said. "That is up to you. But it does seem possible that through your many business and professional associations you *do* have channels of communication with the Hill—the Bureau of the Budget, and even into the Smithsonian administration. Any way that can be found to call attention to the situation of the Air and Space Museum and to create some favorable reactions when museum dollars and personnel allotments are under consideration will be all to the good."

THE GILBERT ARTICLE AND BARRY GOLDWATER

The lack of a proper building to house the collection had begun to attract attention even before Johnston's retirement. In a March 1969 article in *Flying* magazine titled "The Dream and the Junkyard," James Gilbert told his readers about the deplorable condition of the collection at the National Air and Space Museum, especially at the Silver Hill facility. Gilbert began the article by talking about James Smithson's bequest, and how the Smithsonian Institution had become over the years a place for collecting airplanes and related objects, including "the Wright Brothers' very first; and Hiroshima's nemesis, the *Enola Gay;* around 300 aero engines; innumerable models; and not a few NASA spacecraft." Gilbert then asked, "What is the Smithsonian doing with these aeronautical treasures? Not much, if the truth be known."

OPPOSITE: Concepts for the National Air and Space Museum evolved through the years: (top to bottom) the McKim, Mead and White 1955 design; a model of the 1955 design that includes the Smithsonian Castle at left; the Hellmuth, Obata + Kassabaum (HOK) 1964 design; and the HOK 1972 design that became the National Mall building.

COOKIE JAR 1962–68 Shaped like the Mercury capsule, this and many other items were produced by McCoy Pottery for the popular market during the Space Race.

Gilbert complained about the difficulty of locating aeronautical artifacts at the Smithsonian Institution: "Just *finding* the National Air and Space Museum is hard enough. You must start by looking in a gloomy prison of a structure beside Washington's mall. Its wall bears, besides the efflux of an air-conditioner that more truly belongs on display inside, the legend *Smithsonian Institution National Museum 1879, Arts and Industries Building,* which panel has perhaps been there since the building opened, and makes no mention of air or space. You enter upon a dingy, flagged hall, nearly empty, except overhead, where hang the Wright Flyer and the *Spirit of St. Louis."* Gilbert noted that the aeronautical artifacts and exhibits in the Arts and Industries Building and the other building reserved for aircraft, the Tin Shed, were scattered, illogically arranged, out of date, and often unidentified.

But Gilbert reserved the worst of his acid pen for the museum's Silver Hill facility. "What we have seen," he said, "is only the peak of the iceberg; the bulk of it lies underwater, almost literally, for most of the Smithsonian's 207 airplanes repose in what can only be described as a junkyard. . . . This facility is not open to the public; indeed, when we asked a museum official if we could visit Silver Hill, he said, 'only accompanied by someone from museum headquarters, and there's nobody available.' " Gilbert did manage to arrange a tour of Silver Hill, and he concluded that it was "a shame and a disgrace."

Gilbert noted that what was inside the buildings was "bearable." "But the real disgrace," he wrote, "is what is *not,* for lying about the facility are dozens of unique and priceless airplanes simply exposed to the weather to rot. And rotting they are, for though some half-hearted attempts have been made to cocoon their more vulnerable parts, the birds and perhaps other vandals have pecked great holes in the plastic." Gilbert enumerated the items he saw exposed to the elements: a Douglas DC-3, called the *Que Sera Sera,* the first airplane to land at the South Pole; President Truman's presidential airplane; a Douglas C-54, the *Sacred Cow;* World War II–era Grumman fighters; the Ryan Fireball; a Junkers Ju 388; an early Sikorsky amphibian; and a McDonnell F-4, among others. Gilbert decided that "after seeing how it cares for its legacies, I wouldn't donate the museum the ferrule off my shoelace."

Gilbert concluded that dreaming about the museum's future "should not prevent the Air Museum staff from at least labeling what they have managed to put on display now. The Air Museum people seem to have been expecting work to start on their new building every day for 20 years now, and in consequence, putting off any real work of importance till then." He further suggested that the estimated construction cost of $45 million to $46 million could be pared down and that the museum should look to other efforts to build similar museums. "The Smithsonian's justifications for its inability to exhibit and care for its aviation treasures," Gilbert wrote, would be more acceptable were it not for the shining examples of others in the field. The London Science Museum, equally bedeviled by lack of funds, has built a masterly display of its own National Aeronautical Collection." Gilbert cited other examples of museums that had | to page 219 |

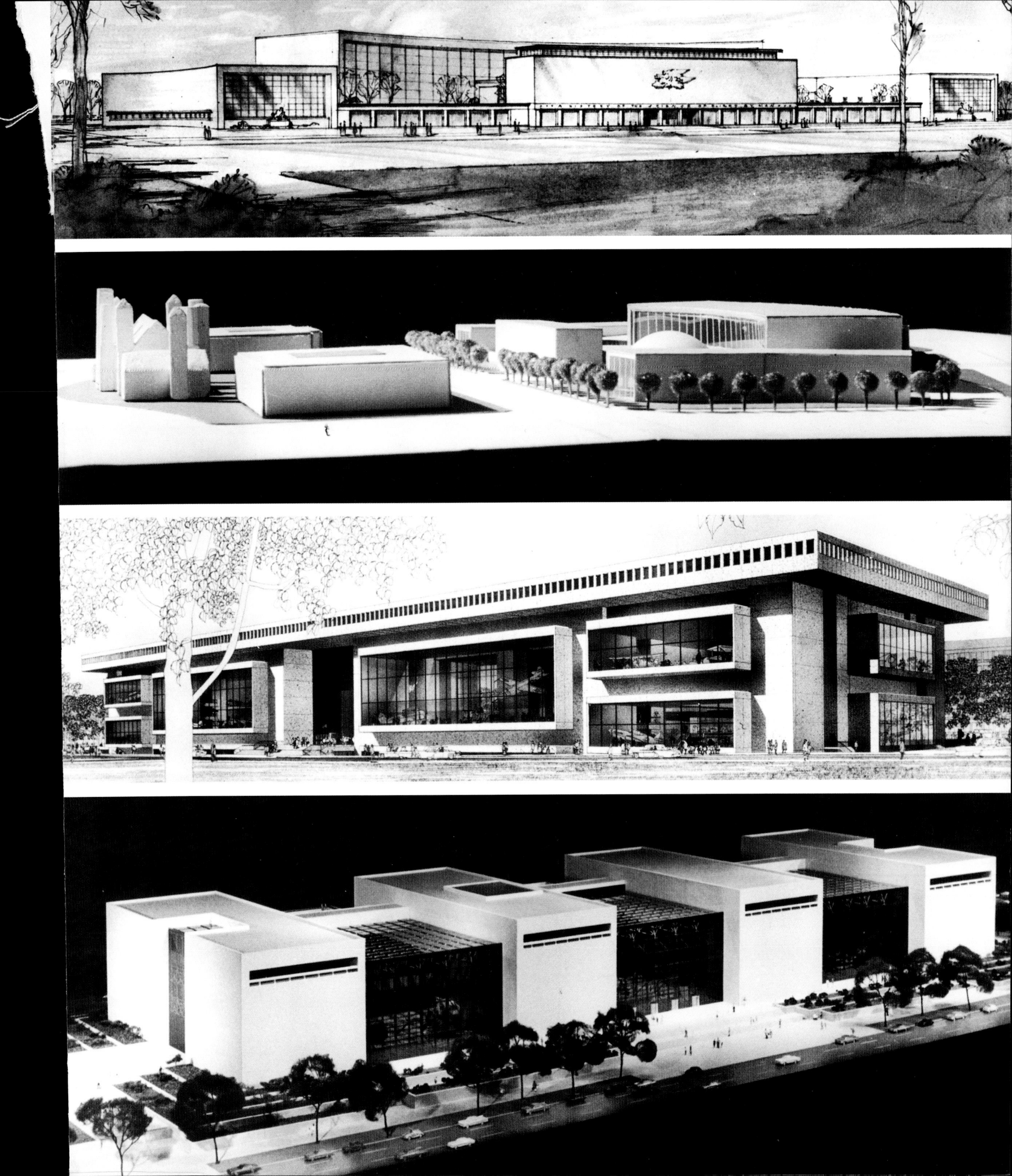

In this front-view architectural model of the center section of the 1955 design proposed by McKim, Mead and White for the National Air Museum, aircraft models hang inside, representing the appearance of the museum with some of its collection displayed.

THE JAM
MEMORIAL

At the August 1969 retirement party held in Arts and Industries for Director S. Paul Johnston, attendees encircled an architectural model of the 1964 museum design concept. Johnston is at center, wearing a bow tie. Present are (left to right) an unidentified woman, Louis Casey, Paul Garber, Kenneth Newland, James Mahoney, E. W. Robischon, an unidentified man, Frederick Durant, and another unidentified man.

| *from page 214* | managed to display their aeronautical treasures under similar financial difficulties, such as the French Musée de l'Air and the Air Force Museum in Dayton, Ohio.

In his speech to the Aero Club, S. Paul Johnston had countered Gilbert by saying that he had exaggerated the situation. "The basic problem does not lie in the area of assisting the museum to salvage a lot of old aircraft which are *allegedly* rotting away on a Maryland hill-top. Actually, of the 200-odd airframes and the 300–400 engines, etc. at Silver Hill—only a relatively small fraction are of sufficient historical value to even be considered for public display. Our physical problems have long since been recognized, and are being dealt with as rapidly as our capabilities (money and manpower) allow."

Moreover, Gilbert had interviewed Johnston for the *Flying* piece, and Johnston had told him that the museum had been constricted by a very small staff and a lack of funds. Not only that, but abundant funds would be necessary to meet the museum's and the Smithsonian Institution's exhibition goals. "An adequate museum today," Johnston told Gilbert, "has got to explain, what does this mean? Our museum is not a static collection of objects, but some kind of dynamic presentation, so that something rubs off on people as they go through. Meanwhile there's a tremendous lot to be done; we are constantly looking at different kinds of display techniques."

James Mahoney, Jr., who was assistant director for exhibits at the museum from 1966 to 1969, and who had been working with the NAM on exhibit problems since 1959, agreed that Gilbert was guilty of | *to page 222* |

James Dean was the first curator of art at the National Air and Space Museum, serving from 1974 to 1980. From 1962 to 1974, he had been the first director of the NASA Art Program. When Dean came to the museum, he brought the bulk of the NASA collection with him, and it has become the backbone of the museum's collection. These thoughts come from his article "The Artist and Space," which appeared in *Interdisciplinary Science Reviews* in 1978.

JAMES DEAN

ON THE NASA ART PROGRAM

Apollo 11 *left the Earth on 16 July 1969 . . . Among those present . . . were a group of artists who interpreted this historical event in their own way: from the suiting-up of the astronaut crew to the launch of the Saturn 5 rocket. Artists were at the Houston Control Center during the landing on the Moon and on the carrier in the Pacific Ocean for the splashdown. The documentary art produced during this historic flight—along with many other drawings, paintings, and prints commissioned by . . . NASA . . . provide a unique record of man's activities in developing his ability to fly through air and space. . . .*

Some may wonder why paintings and drawings of such well-photographed events were needed at all because when a major launch, or test flight, took place, NASA's cameras recorded everything. . . . I am convinced that artists should be key witnesses to history in the making, and that in the long run the truth seen by an artist is more meaningful than any other type of record. . . . I want to build up a collection of drawings and paintings which will convey to future generations some of the excitement and wonder which we feel as we cross the space threshold. I also want these paintings to stand on their own merit as works of art, quite apart from the subject matter. I hope future generations will realize that we have not only the scientists and engineers capable of shaping the destiny of our age, but artists worthy to keep them company. . . .

Most [artists] chose Cape Canaveral for their first assignment. Even in the early days of the Space Age the shapes of structures at this launching site were more fantastic than any artist could imagine. The red-painted launch towers holding the white rockets and the bee hive-like control centers, where engineers—safe under several layers of concrete-filled sandbags—watched their charge through periscopes, were objects of instant fascination to the artists. . . .

Feelings of satisfaction and accomplishment have come with the two major exhibitions of space art at the National Gallery of Art [in 1965 and 1970] . . . and three exhibitions that toured throughout the United States and Europe. Most recently, the rewards have come from the exhibition of this art in the new National Air and Space Museum of the Smithsonian Institution, and watching the public reaction to the work.

| from page 219 |
exaggeration. Mahoney's job was to establish the presence of the National Air Museum because visitors were not aware of it as an entity on the mall, only that the Smithsonian "owned" the Wright Flyer and the *Spirit of St. Louis.* Mahoney began by creating a logo and standardized forms and letterhead for the museum as a way of establishing its identity. Next, he worked to make the public aware of the aircraft in the Tin Shed and to redesign the exhibit spaces for aircraft and establish a context for them. Thus, the Wright EX *Vin Fiz* was heralded for making the first long-distance flight across the United States, the Lockheed 5C Vega *Winnie Mae* was lauded for making record-breaking high-altitude flights, and the Bell X-1 *Glamorous Glennis* was recognized as a pioneering supersonic aircraft. Mahoney also had jurisdiction over the preservation and restoration department at Silver Hill. He confirms what Johnston had contended in his Aero Club speech—that much work had been done by the staff to ensure that as many aircraft as possible were placed under protective cover, but that a shortage of personnel and money had prevented the museum from going any further.

From all indications, Gilbert's article was bombastic and unfair, but it had an effect on the aerospace community, as did S. Paul Johnston's remarks to the Aero Club of Washington. We do not know for certain if Gilbert's *Flying* magazine article was read by Barry Goldwater, the powerful and respected Republican senator from Arizona. We do know that Johnston's well-known speech caught Goldwater's attention. Goldwater was a longtime pilot and aviation advocate, having served in World War II as a reserve officer and Air Transport Command pilot in the U.S. Army Air Forces, delivering aircraft and supplies to overseas destinations. Goldwater had spent a good deal of the war flying between the United States and India via the Azores, North Africa, South America, Nigeria, and central Africa. He also had flown "the Hump" over the Himalaya to deliver supplies to China. He remained in the U.S. Air Force Reserve after the war, eventually retiring as a major general.

MERCURY CAPSULE ICON Ca 1960 Moved along the tracking map of the Mission Control Center at Cape Canaveral, Florida, during early missions

OPPOSITE, TOP: Alan Shepard (second from left) is greeted by fellow astronauts "Deke" Slayton (far left) and "Gus" Grissom (far right) at Grand Bahama Island after his 15-minute Mercury suborbital flight on May 5, 1961. The man in Bermuda shorts is Dr. Keith Lindell, head of astronaut training.

OPPOSITE, BOTTOM: John Glenn makes history on February 20, 1962, inside his *Friendship 7* spacecraft during the first American orbital flight.

Early in 1970, Goldwater had made some preliminary remarks about the plight of the museum to the National Press Club and in a short speech to the U.S. Senate. On May 19, 1970, in a major address on the floor of the U.S. Senate, Goldwater cited Johnston's address to the Aero Club of Washington and criticized Smithsonian officials for not supporting the construction of a new building to house the collections of the museum. Then he began a long critique of the Smithsonian, beginning with the decision to place the National Air and Space Museum under the purview of the assistant secretary for History and Art. "Aeronautics and space exploration," Goldwater said, "derive from, and indeed incorporate, many of the sciences, including mathematics, physics, fuel chemistry, metallurgy, physiology, psychology, biology, astronomy, astrophysics, geology, and geophysics. How in creation flight ever got mixed up with the 'arts' at the Institution is beyond me."

Like Johnston, Goldwater attacked the Smithsonian Institution's administration for withholding funds from the National Air and Space Museum's budget, especially since the museum was "the center in the United States for the exhibition of the historical mainstream of flight" and the "center for public education or awareness of the significance and scope of aeronautical and astronautical developments." Goldwater observed that the museum was expected "to provide reference services for technical historians, patent researchers, airplane building hobbyists, and other specialists who need access to the original sources, including both research materials and actual artifacts."

The National Air and Space Museum had relatively few employees, considering the scope of its responsibilities, and it received only a fraction of the entire Smithsonian budget. "The actual number of personnel at the National Air and Space Museum," he said, "is 30 people, counting the administrative side, support crew, and secretaries. This number also includes the 14 employees working at Silver Hill. . . . In terms of the overall Smithsonian picture, the Air and Space Museum has 1.6 percent of the actual total employment." In terms of budget, he said that "when we examine the money appropriated to the museum—$664,000—in light of the total money appropriated for Smithsonian salaries and expenses—$29,565,000—in the current fiscal year, we come up with almost the same low percentage as resulted in the case of employment."

Goldwater concluded by saying, "I do not know of any better investment for the Nation's celebration of its 200th anniversary than the establishment of a permanent building for the National Air and Space Museum. Aviation and astronautics are America's triumph. Let us recognize it."

After his May speech to the Senate, Goldwater continued to draw attention to the plight of the museum, appearing before the House Subcommittee

on Libraries and Memorials on July 21, 1970. This time, he noted that although Johnston had given the Smithsonian administration 13 months' notice of his retirement, "the Museum is still without a director." "This fact alone" can be interpreted as a lack of interest by the Smithsonian in its Air and Space Museum," he said.

In response to Goldwater's critical remarks, S. Dillon Ripley, the Secretary of the Smithsonian Institution, wrote Goldwater a six-page letter on June 5, 1970, in which he politely defended the Institution and himself concerning the delays in getting an appropriation for the construction of the National Air and Space Museum and the role of science at the Smithsonian. Ripley noted that requests for an appropriation for construction in fiscal year 1966 and 1967 for an estimated $40,045,000 and $40,331,000 had been deleted by the General Services Administration before they even got to the Bureau of the Budget. Ripley also pointed out that in a spirit of compromise, for the 1968 fiscal year budget, the Smithsonian had "decided that an incremental request for construction funds for the foundation and underground parking garage might be more acceptable, following the precedent used by the Public Buildings Service of the General Services Administration to start the FBI Building and the new Labor Department Building. Funds in the amount of $9,500,000 were therefore requested for this purpose in both the fiscal year 1968 and 1969 budget submitted to the Bureau of the Budget and each time the item was deleted and not submitted to the Congress."

As for the Smithsonian Institution's reputed neglect of science, Ripley pointed out that although institutional support for the sciences from the National Science Foundation had declined, the administration had convinced Congress "of a direct appropriation for research. . . . Since that time, Congress has continued to appropriate funds for scientific research, but unfortunately the level of support has remained static. Nonetheless, the largest percentage of funds from that appropriation has been awarded to scientists in the National Museum of Natural History." Ripley went on to say that funds for scientific research at the Institution had been supplemented by grants from other agencies and "from the Institution's limited private resources."

The debate about how much institutional support there was for the construction of a National Air and Space Museum building, the placement of the museum within the purview of history and art, and the Institution's support of research in the sciences appears to have a larger context. In a document titled "A New Look for NASM, Horizons Unlimited for 20th Century America," dated May 1967, S. Paul Johnston laid out some new directions for the museum, including an emphasis on thematic ideas instead of the tried-and-true exhibitions halls of this and that. Johnston wrote that it was time to jettison "the fetish of the famous firsts" and focus on the "potential of man-flight" rather than the "enshrinement of evidences of past accomplishments." "We must therefore," Johnston wrote, "abandon conventional ideas of Halls of Balloons, Halls of Airships, Halls of World War I. . . . The specimens on display should serve to illustrate the ingenious ways whereby man has utilized his growing technological capability in air and space to attain social, political, and economic goals."

These were laudable goals indeed, but how much they reflected Johnston's own thinking, as opposed to his desire to satisfy Secretary S. Dillon Ripley's avowed aims for the Smithsonian, is anybody's guess. Ripley apparently had a low opinion of Johnston and his staff, which included no Ph.D. historians or other scholarly specialists. In a June 8, 1967, memo to his staff, for example, Ripley had argued that if the museum did not take steps to ensure that its employees had adequate academic qualifications and to create a credentialed department of the history of air and space within the museum, he would never be able to secure congressional consent to construct the museum building. Ripley believed that without "consultation from universities, from other museum people in the Smithsonian, and with related groups," the Smithsonian would be unsuccessful in its attempt to create "a meaningful center for research in air and space." The implication was that Ripley thought the existing staff was inadequate to accomplish these goals.

OPPOSITE: An Atlas ICBM launches John Glenn into orbit from Cape Canaveral, Florida, in February 1962.

WHITE SILK VEST 1970
Sewn by the wife of Flight Director Eugene F. "Gene" Kranz for him to wear during the Apollo 13 mission

THE COLLINS ADMINISTRATION AND CONSTRUCTION

As a result of his dissatisfaction with the director and staff of the National Air and Space Museum, Ripley was determined to manage the selection process for the next director himself. He and the search committee, which included Assistant Secretary for History and Art Charles Blitzer and Assistant Secretary for Science Sidney Galler, began to draw up a set of desired characteristics that differed markedly from the ones used to select former directors Hopkins and Johnston. Ripley and the selection committee wanted someone who was intimately familiar with aerospace, well connected to the industry and the military, administratively competent, but above all, innovative.

The selection committee presented several candidates as acceptable, but Ripley dragged his feet on making a decision. In August 1970, he had concluded that "there does not seem to be a really good candidate . . . in sight." In October 1970, the search committee sent letters to prospective candidates to inform them that the current search was being closed and that a completely new search was expected in the future. Apparently, Ripley was troubled by the preponderance of military candidates. In response to the recommendation of a particular admiral, Ripley wrote: "Our principal problem in searching for a Director for the National Air and Space Museum has come in connection with candidates who were in the military service. We have had a sort of rule of thumb in the search for a Director that we would hope to avoid recruiting a military officer as there are museums to do with military aircraft elsewhere in the country and in the world and we are anxious to maintain the image of the National Air and Space Museum as essentially civil."

Ripley also apparently realized that Barry Goldwater had not been appeased by his letter defending himself and the Smithsonian. Goldwater was relentless in his behind-the-scenes advocacy of the National Air and Space Museum. After his major address to the Senate on May 1970, Goldwater directed two

members of his staff, his administrative assistant and his legal counsel, to meet with high Smithsonian officials to review the complaints of neglect and to come up with a solution. In July 1970, the House Subcommittee on Libraries and Memorials began the first legislative oversight hearings on the Smithsonian in more than a century.

On July 22, 1970, Goldwater gave his testimony, emphasizing the pressing needs of both the National Air and Space Museum and the Museum of Natural History. Goldwater emphasized that neither museum had sufficient staff or budget to carry out its assigned responsibilities to the public. In August 1970 Goldwater made another speech to the Senate in which he presented evidence that support for science at the Smithsonian had been reduced. Responding to these pressures from Goldwater, the Smithsonian gave the National Air and Space Museum authority to hire more staff to bring its number of employees up to 41.

In January 1971 Ripley extended an invitation to Goldwater to co-chair and serve as moderator of a special meeting with Smithsonian and museum staff members to discuss what size building should be sought from Congress. The group unanimously agreed that the proposed museum should be redesigned and scaled down in cost from the original estimate of $70 million to $40 million. When the Smithsonian Institution's budget was presented to Congress in 1971, it included a request for $1.9 million to redesign the museum building and specified that a deadline for completion be set for July 4, 1976. Goldwater then testified on March 22, 1971, before the Senate Appropriations Committee in support of the redesign and the construction deadline. Goldwater subsequently made other official pleas on behalf of the museum, including one to President Nixon after the Bureau of the Budget had again rejected the museum's request, in which he persuaded the President to overrule his budget officials. In August 1971 Congress passed a funding measure for $40 million toward the new building. It had earlier appropriated the $1.9 million for Hellmuth, Obata + Kassabaum to redesign the building, which had to be reduced in volume by half. Thus the path was finally cleared for the construction of a permanent home for the National Air and Space Museum.

Meanwhile, after more than a year and a half since Johnston's bitter departure in September 1969, Ripley hired a new museum director in April 1971, former Apollo 11 astronaut Michael Collins. Collins had left the astronaut corps and NASA and had been the State Department's assistant secretary of state for public affairs. Largely forgotten was that he had been a crack USAF test pilot and that he still held the rank of colonel. Here, then, was a career military man, but with a distinct difference. Collins met all of the criteria set forth by Ripley in the first iteration of the search: He was knowledgeable about aerospace, well-connected to the industry and the military, and administratively savvy. For good measure, he was a celebrity and an American hero, having taken part in the historic voyage to the moon in July 1969. The obvious question on everyone's mind was: Would he be innovative? | *to page 232* |

The museum's signature artifacts formed the central display in the Arts and Industries Building in May 1973: the Wright 1903 Flyer (top foreground), Ryan NYP *Spirit of St. Louis* (hanging in the rear), the Gemini IV spacecraft (at left), the Apollo 11 command module *Columbia* (right), and lunar module 2 (behind *Columbia*).

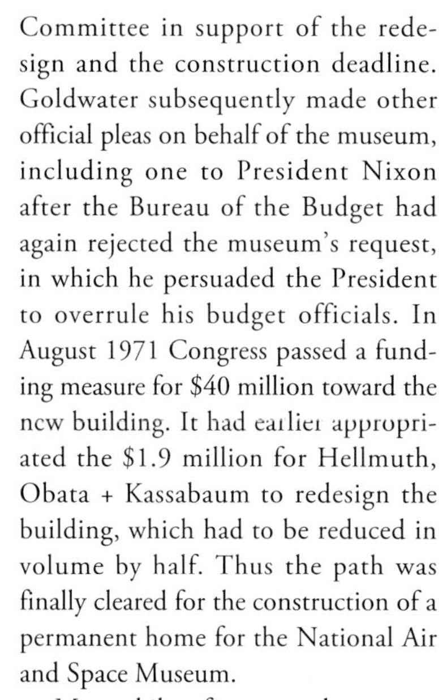

U.S. AIR FORCE PILOT'S NAVIGATION KIT 1964–73 Used by F-105 fighter-bomber pilot Capt. Ben Allen in missions over North Vietnam

THE ORIGINAL WRIGHT BROTHERS
AEROPLANE, 1903
THE ORIGINAL
WRIGHT
BROTHERS
AEROPLANE
Spirit of St. Louis
APOLLO 11
COMMAND MODULE

Alan Shepard's capsule was displayed in the museum's Aircraft Building (also called the Tin Shed) in the early 1960s.

On May 5, 1961, NASA astronaut and Navy Cdr. Alan B. Shepard, Jr., became the first American, and the second human, in space. Lobbed on a suborbital trajectory by a Redstone rocket, Shepard and *Freedom 7* reached a height of 116.5 miles before splashing down 302 miles away in the Atlantic. The mission, officially called Mercury-Redstone 3, lasted 15 minutes and 28 seconds, and came after the Soviet Union launched Yuri Gagarin into orbit on April 12.

MERCURY CAPSULE

FREEDOM 7

Project Mercury arose in 1958 when the United States found itself in a Space Race. President Dwight Eisenhower gave the prized human spaceflight assignment to the new civilian agency, NASA. At the Langley Research Center in Virginia, brilliant engineer Max Faget and his collaborators designed the capsule; McDonnell Aircraft Corporation of St. Louis, Missouri, won the development and production contract. Mercury's objective was to place an astronaut in orbit using an Atlas ICBM, but NASA felt it wise to launch test flights on the smaller Redstone missile.

Alan Shepard in *Freedom 7* prepares for his historic 15-minute suborbital ride. He became the first American, and the second human, to fly in space. OPPOSITE: Shepard on the Navy destroyer U.S.S. *Lake Champlain,* along with his capsule, after splashing down in the Atlantic.

Shepard was selected as one of the first seven astronauts in May 1959. His spacecraft, only six feet in diameter, was so snug that the astronauts joked that "you didn't get into it, you put it on." Weighing just over 3,000 pounds, it discarded its retro-rocket package and parachute container before landing. Shepard named his spacecraft *Freedom* and added the numeral seven for the number of astronauts.

The Smithsonian unveiled *Freedom 7* in the Arts and Industries Building on October 23, 1961. It long had a place of honor, but the spacecraft, like Shepard's mission, was ultimately displaced in the American mind by *Friendship 7,* as many felt that John Glenn's 1962 orbital flight was the one that equaled Gagarin's achievement. It is Glenn's capsule that sits in the museum's Milestones of Flight gallery today.

| from page 226 |

On August 24, 1971, Collins sent a memo to the staff entitled "What Do You Need in the New Building?" Collins pointed out that despite the earlier, more ambitious plans for the museum, "Our target is to complete construction in July 1975 (to allow time to set up exhibits properly for a July 4, 1976 opening) and for the total project cost (excluding exhibits) not to exceed $41.9 million. . . . I think it is essential for each of us on the staff to wrack our brains now and unearth each last little detail that needs to be built into the new building. Generally speaking such details cost little or nothing now but become increasingly expensive as the building nears completion. Examples are special requirements for electrical power, humidity control, sound equipment, noise control, disappearing screens, floor loading, built-in display cases, moveable walls, etc. Not only should you consider your exhibition requirements, but those for office space and workshops as well, especially if you have unconventional needs."

In addition to the details of the building, Collins wanted to make sure that he had a team capable of creating the museum and filling it with top-notch exhibits. He started by hiring Melvin Zisfein as deputy director. Zisfein held a Ph.D. in aeronautical engineering from MIT and had worked in advanced projects at Bell Aircraft before heading the research laboratories at the Franklin Institute in his native Philadelphia. A brilliant engineer, Zisfein also had a taste, and a talent, for communicating complex ideas to the public.

The new director also created a Department of Science and Technology to broaden the professional base of the staff. It was headed by Howard Wolko, an authority on aircraft structures with a long record of work both in industry and academic engineering, who attracted several young doctoral candidates in the history of science and technology who would play important roles in planning the museum.

Collins took another major step toward the professionalization of the museum staff when he annexed personnel and critical elements of the Bellcomm, Inc., operation in 1972–73, when the former NASA contractor was being closed down. Farouk El-Baz was hired to lead the museum's new Center for Earth and Planetary Studies. El-Baz had been instrumental in selecting lunar landing sites for the Apollo Program. Under his direction, CEPS became a unit devoted to active research in analysis of lunar and planetary spacecraft data and the lead center for Earth observations and photography from the Apollo-Soyuz Test Project, in which a Soviet and an American aircraft docked in space in July 1975. El-Baz became so well known in the space community that the producers of *Star Trek: The Next Generation* named one of their shuttlecraft—the *El-Baz*—after him.

CEPS acquired a vast collection of planetary photographs and images from NASA. That collection, which has expanded in the 35 years of its existence, now makes up the bulk of a NASA Regional Planetary Image Facility at the museum. It includes an extensive collection of images of the planets and their satellites, as well as photographs taken by space shuttle astronauts using handheld cameras in addition to some radar imagery.

Another Bellcomm veteran, Donald S. Lopez—a former U.S. Air Force lieutenant colonel with a record as a World War II fighter ace and a master's degree in thermodynamics from the California Institute of Technology—came on board as chairman of the Aeronautics Department in 1972. He would spend more than three decades at the museum, eventually serving as both acting director and deputy director. Catherine "Kitty" Scott, Bellcomm's information specialist, brought most of the company's technical library with her when she came to the museum. Scott created a first-class library at the museum and hired professional staff.

Collins also brought the first professional historians on staff. At the time, professional history at the Smithsonian was centered at the Museum of History and Technology, now called the National Museum of American History. By the 1960s, it had become a national focal point for research and graduate training in the history of science and technology. Collins used the Smithsonian fellowship program to attract fresh talent to the National Air and Space Museum. Many of the scholars who came to pursue research at the museum would remain as employees.

At the time, the Daniel and Florence Guggenheim Foundation had pledged up to a quarter million dollars (an amount roughly equal to $1.7 million in 2010) to support three projects at the National Air and Space Museum. The funds created an exhibit about the Guggenheim family's long-standing support of aeronautical research and education, sponsored an annual public lecture series by distinguished figures from aerospace history, and established | to page 237 |

CHECKLIST DRAWING 1969 Ground crew put this drawing into Michael Collins's command module checklist for Apollo 11, the first moon landing.

OPPOSITE: A contact print of photographs taken at the Great Hall of the Smithsonian Institution shows festivities at lunch after the ground-breaking ceremony for the National Air and Space Museum on November 20, 1972.

KODAK SAFETY FILM
KODAK TRI X PAN FILM
→1A →2 →2A →3 →3A →4 →4A →5 →5A →6
KODAK TRI X PAN FILM
KODAK SAFETY FILM
→7A →8 →8A →9 →9A →10 →10A →11 →11A →12 →12A
FILM
KODAK SAFETY FILM
KODAK TRI X PAN FILM
→14 →14A →15 →15A →16 →16A →17 →17A →18 →18A →19
KODAK SAFETY FILM
KODAK TRI X PAN FILM
KODAK
→20 →20A →21A →22 →22A →23 →23A →24 →24A →25
KODAK TRI X PAN FILM
KODAK SAFETY FILM
→26A →27 →27A →28 →28A →29 →29A →30 →30A →31
TRI X PAN FILM
KODAK SAFETY FILM
KODAK TRI X
→32A →33 →33A →34 →34A →35 →35A →36 →36A

The construction site of the National Air and Space Museum shows work well under way on December 14, 1972.

#16

| from page 232 | annual fellowships to allow graduate students to pursue historical research at the museum. Richard Hallion, a graduate student from the University of Maryland, with a book on the Bell X-1 and the Douglas D-558 already, was awarded a Guggenheim predoctoral fellowship grant to do research on the Daniel Guggenheim Fund for the Promotion of Aeronautics. He was subsequently hired as a staff member.

Several other young and promising graduate students came aboard. Paul Hanle, who was in the process of earning a Ph.D. in the History of Science and Medicine from Yale University, joined Hallion on the staff of the museum's new Department of Science and Technology. Tom Crouch, a Ph.D. candidate from the Ohio State University, arrived in 1973 with a short-term Smithsonian fellowship to pursue his dissertation research in the Smithsonian's Samuel P. Langley papers and at the Library of Congress. Crouch was a veteran of the museum program of the Ohio Historical Society, where he had planned the exhibitions for the Neil Armstrong Museum in the astronaut's hometown of Wapakoneta, Ohio, as well as the history display for the newly opened Ohio Historical Center in 1971. A year after his fellowship, he accepted Collins's offer of a job.

One of the first tasks for the new professional staff was to create the exhibits for the new building. That task fell directly to Melvin Zisfein, who came with fresh ideas about how an aerospace museum could inspire and educate the visitor. He drafted an informal report on his vision for the museum, which he submitted for Secretary S. Dillon Ripley's consideration. The report contained suggestions for interactive exhibits and focused on behind-the-scenes areas of air and space that had never been explored by the museum. Ripley appeared to be pleased with the report, complimenting Zisfein on his "interesting and imaginative ideas." Ripley concluded that the report "certainly demonstrates a dimension of exhibition that we are striving to achieve in improving the educational and communicative roles of the Museum."

Among his many exhibit ideas, Zisfein envisioned a Milestones of Flight gallery, where aircraft and spacecraft that had significant flights and had overcome physical, rather than man-made, obstacles (that is, set basic records like first, fastest, or farthest to fly), would be displayed. Among the artifacts chosen were the Wright Flyer, the Bell X-1 *Glamorous Glennis,* and the Ryan NYP *Spirit of St. Louis,* along with Mercury *Friendship 7,* Gemini IV, and the Apollo 11 command module *Columbia.* Zisfein also came up with the idea of developing and testing "practice" exhibitions in the Arts and Industries Building. Such displays could then be adapted for the new museum.

The pressure created by the need to fill the building changed the exhibit development process in the Smithsonian. Previously, exhibition designers were drawn from a pool of talent maintained by the Office of Exhibits Central. Given the relatively slow pace of exhibition development in those years, it did not make sense for individual museums to support their own dedicated staff of designers. Now the Smithsonian administration was forced to rethink that traditional strategy. Although Exhibits Central would continue to function, designers were reassigned to individual museums.

COMMUNICATIONS CARRIER
1969
Worn by Edwin E. "Buzz" Aldrin, Jr., during the Apollo 11 moon landing in July 1969 and nicknamed a Snoopy cap

OPPOSITE:
A sequence of views of the construction of the National Air and Space Museum taken in 1974, approximately 18 months after ground had been broken

At the National Air and Space Museum, this meant the creation of an Exhibits Department, complete with designers, production craftspeople, and for a time, professional illustrators. At the outset, James Mahoney, who had become head of Exhibits Central, oversaw exhibit development at the museum. Once its development program was under way, Mike Collins hired Tony Baby to supervise design and production activities. Hernan Otano, a colorful veteran of the Argentine Air Force, was responsible for electro-mechanical interactives.

Even with these changes, museum leaders recognized that the curatorial and design staffs on hand could not fill the new building with displays without considerable outside assistance. The standard practice during the period 1972–76—when the exhibitions for the museum were conceived, scripted, designed, and produced—was for a staff curator and designer to work with a contract exhibit firm to create a single gallery.

Mike Collins was especially concerned about the museum's role in fostering flight and the arts. During the NAM period, art had been included in the collections, but not in a formal way. The S. Paul Johnston administration, in its 1965 report "Proposed Objectives and Plans for the National Air and Space Museum," had recommended that because of "the richness of subject matter, its historical interest, and its cultural values, . . . adequate provision should be made in the new facility for permanent art displays as well as for rotating or 'loan' exhibits." Collins now had the resources to take that advice seriously. He hired James Dean, an artist who had played a key role in creating and managing NASA's art program. When Dean moved to the new building, across the street from the space agency, he brought the entire NASA art collection with him, instantly transforming the museum into one of the world's great repositories of aerospace art.

To show off some of the gems of the now greatly expanded collection, Collins envisioned a Flight and the Arts gallery in his plans for the building. He also made arrangements with artists Eric Sloane, who created "Earth Flight Environment" (on the west side), and Robert T. McCall, who did "The Space Mural—A Cosmic View" (on the east side), to paint large murals in the south entranceway to the building. Sculptors Charles B. Perry ("Continuum," south entrance) and Richard Lippold ("Ad Astra," north entrance) were commissioned to create appropriate sculpture outside.

MOVING IN

Architect Gyo Obata's goal was to design a building in harmony with the character of the National Mall, and one that would reflect architectural elements of the National Gallery, its neighbor to the north. The finished building was 685 feet long, 225 feet wide, and 85 feet high, with a total area of 630,000 square feet. It was made up of seven sections: Four sections were covered in a pink-hued Tennessee marble, and these alternated with glass sections in three recessed exhibit bays that had | *to page 243* |

The Douglas DC-3 being installed in the Air Transportation gallery in 1975. While the aircraft was being hung, it dropped and one of its wingtips was damaged. After repairs, the aircraft was again hoisted into place.

N18124

The Hawker-Siddeley XV-6A Kestrel is delivered at the west end of the National Air and Space Museum building. The huge doors at that end of the building could be opened and closed to allow large artifacts to be moved in and out with relative ease.

UNITED
& HAULING
937-8510
WIDE LOAD

The Milestones of Flight gallery during installation of artifacts. The 1903 Wright Flyer hangs at center. Behind and above it is the North American X-15 rocket-powered research aircraft. At left is the Gemini IV spacecraft.

| from page 238 | acrylic plastic domes overhead and heavy truss systems to support the weight of aircraft and spacecraft suspended above. The high ceilings were also designed so that rockets and missiles could be placed in a recess in the floor of Space Hall and stand erect.

On September 11, 1972, bulldozers began clearing the museum site, and on November 20 of that year, Chief Justice of the Supreme Court and Smithsonian Institution Regent Warren Burger and Secretary of the Smithsonian S. Dillon Ripley officiated at the ground-breaking ceremony. For the staff, the real work started early in 1975, when they moved into their new building and began to fill it with air- and spacecraft and the exhibits that would surround them. The effort to restore those flying machines had been under way at Silver Hill for some years. The time had come to begin transporting them into Washington from the Maryland suburbs and moving them into the museum through the huge doors in the west end of the building.

The largest single craft to be displayed, a backup Skylab Orbital Workshop, presented special problems. Standing 48 feet tall, measuring more than 21 feet in diameter, and weighing some 78,000 pounds, the spacecraft had been outfitted for flight, but was placed in storage at NASA's Marshall Space Flight Center in Huntsville, Alabama, when the program came to an end. NASA workers carefully cut the craft into three sections and otherwise modified it so that visitors would be able to walk through the living quarters of the first American space station. The sections were then shipped by water to the Washington Navy Yard, transported to the museum, and reassembled in place. Nervous staffers breathed an enormous sigh of | to page 246 |

Rocket engineer Wernher von Braun (left) speaks with his old friend Fred Durant, the head of the Astronautics Department, in the Milestones of Flight gallery in April 1976. Above them hangs the engineering model for the Mariner 2 probe to Venus. OPPOSITE: Artist Robert McCall works on "The Space Mural—A Cosmic View" in 1975.

| *from page 243* | relief when the bolts fell into place.

Howard Wolko, a structural engineer, was responsible for calculating the loads on the museum structure caused by hanging air- and spacecraft. Walter Boyne, a retired Air Force colonel who eventually became director, supervised the process of suspending the artifacts in the various galleries. The Silver Hill crew, famous for their expertise in restoring historic flight craft, did the lion's share of the work as one machine after another was hoisted into place.

The work crew responsible for placing air- and spacecraft "dropped" only one airplane. As the Douglas DC-3 was being lifted into place, clamps on an overhead beam shifted and the airplane swung toward the windows of the Air Transportation Hall, scraping a wingtip across the floor. The DC-3 installation came to an abrupt halt just short of a real disaster. A replacement wingtip was soon on its way from the salvage yard.

By the spring of 1976, the museum was a beehive of activity. Veteran staffers remember the aging artist Eric Sloane, working high on scaffolding, painting the huge mural on the west wall of the south lobby while his assistant kept him entertained by playing a concertina. Robert McCall, who was working on the space mural on the opposite wall, invited astronaut and artist Alan Bean to paint a single star high up near the rooftop domes. Those who were not working on a particular gallery could be found painting the walls, polishing the brass handrails, or performing whatever tasks were necessary to prepare the new museum for the visitors to come. Creating the museum was a team effort in every sense of the word. Now it only remained to see what the public would have to say.

Museum Director Michael Collins sits on the edge of the escalator outside the NASM Theater, finding a moment of quiet during the festivities for the opening of the museum on July 1, 1976.

5

The insignia collection of the museum is one of the largest artifact assortments in its possession. There are some 7,000 individual pieces from more than 65 countries, primarily from the military services and from commercial airlines.

INSIGNIA

REGIA AERONAUTICA **1943**

SOVIET AIR FORCE **1970s** | SOVIET AIR FORCE **1970s** | AMERICAN AIRLINES **1930s** | ALL AMERICAN AIRWAYS **1955** | AMERICAN AIRLINES **1980s**

U.S. NAVY AIR CREW **1945** | AMERICAN INTER ISLAND **1990s** | ROYAL CANADIAN AIR FORCE **1944** | U.S. ARMY AIR FORCES **1944** | ROYAL NETH'DS ARMY **1942**

ROYAL MOROCCAN A.F. **1960s** | PENNA. CENTRAL AIRLINES **1935** | BRAZILIAN AIR FORCE **1970s** | NAVY TACTICAL OBSERVER **1944** | ROYAL NETH'DS AERO CLUB **1930s**

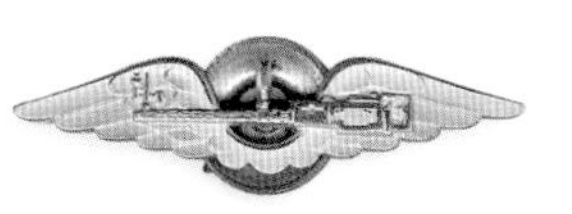

PORTUGUESE AIR FORCE **1960s** | JAPAN NAVAL SELF-DEF. **1970s** | ROYAL SWEDISH AIR FORCE **1945** | ROYAL SWEDISH AIR FORCE **1944** | U.S. NAVY AVIATOR **1970s**

BRAZILIAN AIR FORCE **1970s** | COLOMBIAN AIR FORCE **1980s** | NATIONALIST CHINA A.F. **1970s** | ARGENTINE NAVY AVIATOR **1970s** | U.S. ARMY A.F. CADET **1943**

LOFTEIDIR (IAL) **1980s** | OLYMPIC AIRWAYS **1970s** | ROYAL A.F. AMBULANCE **1944** | BRIT. ARMY PARATROOPER **1944** | ROYAL THAI NAVY **1970s**

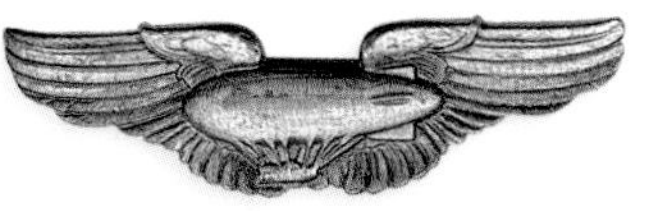

SEABOARD & WEST'N KIDDIE **1950s** | SENEGAL A.F. **1970s** | U.S. ARMY AVIATOR **1970s** | U.S. NAVY ASTRONAUT **1970s** | U.S. ARMY AIR FORCES **1942**

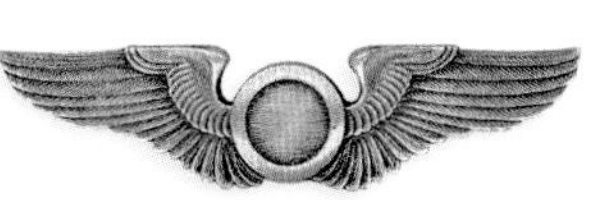
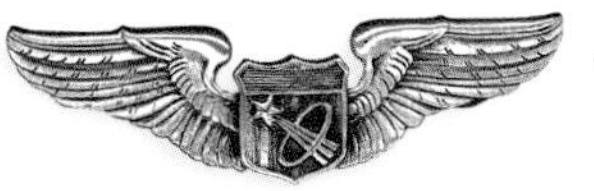

CIVIL AIR PATROL COMM. **1980s** | U.S. ARMY A.F. OBSERVER **1944** | U.S. A.F. ASTRONAUT **1960s** | FINNISH AIR FORCE **1970s** | JAPAN SELF-DEFENSE **1970s**

DELTA AIRLINES 1970s

AIR FRANCE 1950s

CONSOLIDATED AIR LINES 1930s

WOMEN'S VOLUNTARY SVCS. 1943

DELTA FLIGHT CREW 1970s

FRENCH NAVAL AIR SVC. 1970s

CANADIAN AIRWAYS 1930s

AVIANCA 1955

ALASKA STAR AIRWAYS 1930s

BRANIFF AIRWAYS 1930s

REGIA AERONAUTICA 1943

AMERICAN LEGION 1938

AMERICAN AIRLINES 1950s

AIR TRANSPORT COMMAND 1944

ALASKAN AIRLINES 1950s

CIVIL AIR TRANSPORT 1960s

FRENCH AIR FORCE 1970s

ROYAL AIR FORCE 1944

PRAIRIE AIRWAYS 1930s

BOLIVIAN AIR FORCE 1970s

SAUDI ARABIAN AIRLINES 1970s

WESTERN AIR EXPRESS 1930s

WIGGINS AIRWAYS 1930s

SEABOARD & WESTERN 1960

CIVIL AIR PATROL 1985

The military collection comprises rank, unit, rating, and qualification badges. The commercial collection includes the insignia of aircraft crew as well as ground personnel. More than 200 airlines are represented, many of which went out of business decades ago. These badges may be these organizations' only remaining artifacts. / The insignia collection came from many different sources, especially subject-area collectors and individuals donating personal items. The most significant portion of the collection came from the transfer of these artifacts from the Historical Branch of the Institute of the Aeronautical Sciences (IAS), a collection started by aeronautical connoisseur Lester Gardner, whose goal was to emulate the collecting practices of European organizations such as the Royal Aeronautical Society. By the late 1950s, housing the collection in New York became too expensive and the IAS agreed to transfer it to the Smithsonian.

SLICK AIRWAYS **1953**

THE NASA-NASM PARTNERSHIP

The Apollo 4 unmanned mission lifts off from the Kennedy Space Center, Florida, on November 9, 1967. This was the first flight for the enormous Saturn V rocket, which would soon send humans to the moon. LEFT: The pressure vessel for the Variable Density Wind Tunnel arrives at the NACA Langley Laboratory in Virginia in 1922. The Smithsonian played a key role in launching NACA, the predecessor of NASA.

From its very origins, the National Aeronautics and Space Administration has been closely intertwined with the Smithsonian. The fourth Secretary of the Institution, Charles D. Walcott—a geologist now best known for his discovery of the Burgess shale that revealed the "Cambrian explosion" of multicellular life a half billion years ago—played a central role in creating the National Advisory Committee for Aeronautics in 1915 (see Chapter 1). Walcott served to the end of his life as the chairman of the main committee of that collegially run organization; subsequent Secretaries served in leading positions up until the NACA's conversion to NASA on October 1, 1958.

But the aeronautical research institution was not a significant source of artifacts. Until the eve of World War II, it consisted of a small Washington headquarters and the Langley Laboratory (named for the third Secretary) in Hampton, Virginia. It did not create aircraft; it only tested them and their related technologies, notably in its pioneering wind tunnels. The Smithsonian and its National Air Museum received a few airfoils, experimental propellers, and at the very end of NACA's existence, an early sounding rocket.

All that would change when, in its new guise, the agency became an operational entity, charged with building or buying satellites, spacecraft, and launch vehicles. Only two and a half years later, in May 1961, the agency launched the first American into space, Alan Shepard. Not surprisingly, the NAM quickly asked for his Mercury capsule *Freedom 7,* and NASA was happy to oblige, allowing its display in the nation's place of honor.

It was unveiled in the Arts and Industries Building in October. But NASA did not formally transfer ownership at that time, a policy it largely followed for the next several years. While NAM curators treated the capsule as an accessioned object, it was actually on "indefinite loan." NASA's administrators were reluctant to give up control, in part because the agency had its own national and international exhibit program. Mostly that consisted of models and traveling displays, but the exhibit of real spacecraft, especially flown human ones, became an enormous public draw, bolstering NASA's political support. Such artifacts also became useful propaganda tools in the Cold War: The State Department wanted to show them at events such as World's Fairs. Three days after John Glenn became the first American to orbit Earth on February 20, 1962, Smithsonian Secretary Leonard Carmichael asked for that capsule too. NASA officials quickly gave a verbal OK, but Administrator James E. Webb never mailed the affirmative answer drafted for him, as *Friendship 7* was diverted to a world tour. It arrived in Arts and Industries only a year later, again on an "indefinite loan" basis.

All that changed in the mid-1960s. NAM Director S. Paul Johnston and the first head of the Astronautics Department, Fred Durant, worked to develop a close relationship with NASA. Simultaneously, Webb, a fast-talking North Carolina lawyer and consummate Washington insider, became increasingly dissatisfied with "the lack of curatorial expertise by NASA public affairs staff . . . and the lack of protective care in public exhibit of artifacts. He stated that NASA was not in the 'museum business' and was 'not expert in such matters, but the Smithsonian was,' " according to a 1971 memo from Durant. In May 1966, Webb asked the NASA Public Affairs chief to start discussions with Johnston on how to transfer artifacts while sustaining cooperation.

Probably not coincidentally, a bill was moving through Congress at the time to broaden the museum's purview and add "Space" to its name, a change discussed as early as February 1960 in a letter from Secretary Carmichael to NASA's first administrator, T. Keith Glennan. Now there was renewed hope that a building would actually be built on the mall. The upshot was a document signed in March 1967 by Webb, Johnston, and the latest Smithsonian Secretary, S. Dillon Ripley. Informally known as the NASA–NASM Agreement, it gave the Smithsonian a right of first refusal of any historic objects for which NASA no longer had a need. In case the museum decided to deaccession artifacts later, the space agency could choose to take them back. In effect, the National Air and Space Museum became NASA's museum, giving it the inside track on all agency artifacts—a relationship closer even than with the armed services, traditionally the strongest supporters of the NAM. But the services did not have a blanket obligation to offer everything first to the Smithsonian.

Fred Durant and his curators, funded by a sum of money NASA transferred with the agreement, jumped at the chance. They inspected human spacecraft and other artifacts at agency centers and other museums, initiating a massive increase in the number of space-related objects in the National Collection. The first wave came in the year after signing, when the museum moved to take title to the Mercury capsules and whatever Gemini (two-astronaut) spacecraft that NASA would be willing to release, along with important objects of robotic spaceflight. Most of them remained on long-term loan to the NASA visitor centers or other museums where NASA had placed them. Another wave came around 1970–71, when the rest of the artifacts of the Mercury and Gemini program and some early Apollo artifacts were surplused.

Meanwhile, the museum had placed the Gemini IV spacecraft (associated with the first U.S. spacewalk) on exhibit in Arts and Industries in 1966 and the Apollo 11 command module *Columbia* in 1971. Two massive waves of Apollo artifacts arrived in the mid- to late seventies, after Apollo and its successor programs were terminated. Between 1967 and 1980, the museum acquired over 5,000 NASA artifacts, including every spacecraft flown by U.S. astronauts and virtually all of the ones ground tested or flown without pilots. That not only filled the new mall building and the Silver Hill storage facility, it turned the National Air and Space Museum into a major lender, with responsibility for objects all over the world.

This rich panoply of capsules, spacesuits, rockets, and robotic probes was an enormous asset, many of them contributing mightily to the popularity of the mall museum. But it also created long-term problems. Storage conditions at the Garber Facility (as Silver Hill was named in 1980) were often poor, and the environmental conditions of many loan objects were not much better. The museum simply did not have the resources to protect all of its artifacts.

By far the most expensive problems were created by the Saturn V moon rockets that the museum took title to in the mid- to late 1970s. Using surplus hardware from the Apollo program, three NASA

John Glenn's Mercury capsule *Friendship 7* is saluted by a trumpeting elephant at Katunayake Airport, Ceylon (Sri Lanka), July 1, 1962, following a three-day exhibition in Colombo. The capsule now sits in the Milestones of Flight gallery of the museum.

centers—Marshall in Alabama, Kennedy in Florida, and Johnson in Texas—had lined up a collection of stages and real or boilerplate Apollo spacecraft to create horizontal outdoor displays. When stacked vertically on the launch pad, these artifacts—the most gigantic in the Smithsonian collections—would have been 363 feet high. By the mid-1980s, the impact of weathering and animal intrusions was becoming obvious. Veteran rocketry curator Frank Winter visited all three in 1987. He found the Houston one in the worst shape, with holes through the second stage and birds nesting inside.

The first break came in 1993, when the Kennedy Space Center requested permission to restore its rocket and display it at a huge new Apollo/Saturn V Center. The Kennedy Space Center, with its geographic advantages of proximity to central Florida's rising theme parks and a monopoly on space shuttle launches, had turned its Visitor Center into a contractor-operated, fee-charging facility drawing over two million tourists per year. That allowed it to work with local and state authorities to raise money for the project. In 1995–96, supervised by Frank Winter and members of the NASM Collections Division, a Kennedy Space Center contractor carried out a two-million-dollar restoration and moved the pieces into the new multimillion-dollar facility in late 1996.

That success naturally whetted the appetites of curators at the museum and the other two locations. Johnson Space Center in Houston carried out cosmetic and emergency repairs on its vehicle in 1995, but that did no more than forestall even worse damage. As part of a presidentially sponsored Save America's Treasures grant in 1999, the museum won over a million dollars for the Johnson Space Center's Saturn V, the only one made up entirely of flight-qualified stages. But that money had to be matched by private donors—and not all of it ever was. Restoration began in 2004, supervised by curators Allan Needell and Frank Winter, and lasted two years. JSC subsequently found funds to construct a building to protect the precious work of the conservators, leading to a total project of over five million dollars. This Saturn V was unveiled in May 2006.

Meanwhile, the U.S. Space and Rocket Center in Huntsville, Alabama, the visitor center for Marshall Space Flight Center, was able

Work is just beginning on the museum's Saturn V at Kennedy Space Center in January 1996. It was soon tented for restoration. At the end of that year the rocket was moved inside a new display building. OPPOSITE: Restoration work inside the second stage of the Saturn V at NASA Kennedy Space Center in March 1996.

to raise its own money, with state aid, to restore and house the last moon rocket. Work began in 2005. In July 2007 it was moved into the partially completed Davidson Center for Space Exploration. On January 31, 2008, the 50th anniversary of the first U.S. satellite, the USSRC officially opened the new building. Collectively, saving the three moon rockets was an enormous achievement, a partnership not only of NASA and National Air and Space Museum, but also of the visitor centers, state officials, private donors, and contractors.

Another noteworthy case of cooperation has been the shuttle approach-and-landing test orbiter *Enterprise,* which arrived at Dulles Airport in November 1985 on the back of a NASA Boeing 747. It became an artifact crucial to the establishment of the later Udvar-Hazy Center (see Chapter 5). Only a couple of months after the *Enterprise's* arrival, *Challenger* was destroyed in a tragic launch accident. *Enterprise,* originally slated for spaceflight before being replaced by the test article that became *Challenger,* was pressed into service from June to August 1987 to test new safety and escape systems as well as cargo-bay vibration during launch. In 1990, NASA borrowed the main landing gear for use on a NASA landing systems test aircraft; they were returned in 1996.

It was not the first time the museum had lent artifacts to the agency—NACA took a P-61 Black Widow night fighter for testing from 1950 to 1954. After the second space shuttle tragedy, that of *Columbia* in early 2003, *Enterprise* again aided the recovery process. Wing leading-edge panels and the main landing-gear doors were removed for nondestructive testing in spring 2003 and reinstalled three years later, after the *Enterprise* was already on exhibit at the Udvar-Hazy Center.

With the closeout of the shuttle program, the cooperative partnership between NASA and the museum continues. Hundreds of new artifacts will be coming to the museum in the next few years, and Space History curators hope to replace *Enterprise* at the center with *Discovery,* the oldest of the three surviving orbiters. Meanwhile, the space agency has borrowed or analyzed Apollo heat-shield sections, spacesuit parts, and other artifacts as part of engineering its programs. The NASA–NASM Agreement has passed its 40th anniversary and appears to have a bright future. — *Michael J. Neufeld*

NASA workers reinstall the cockpit windows of the space shuttle *Enterprise* in 2007 after they had been borrowed for testing and evaluation in the shuttle program. OPPOSITE: Museum restoration staff complete the cleaning and repainting of the *Enterprise* before the opening of the James S. McDonnell Space Hangar at the Udvar-Hazy Center in November 2004.

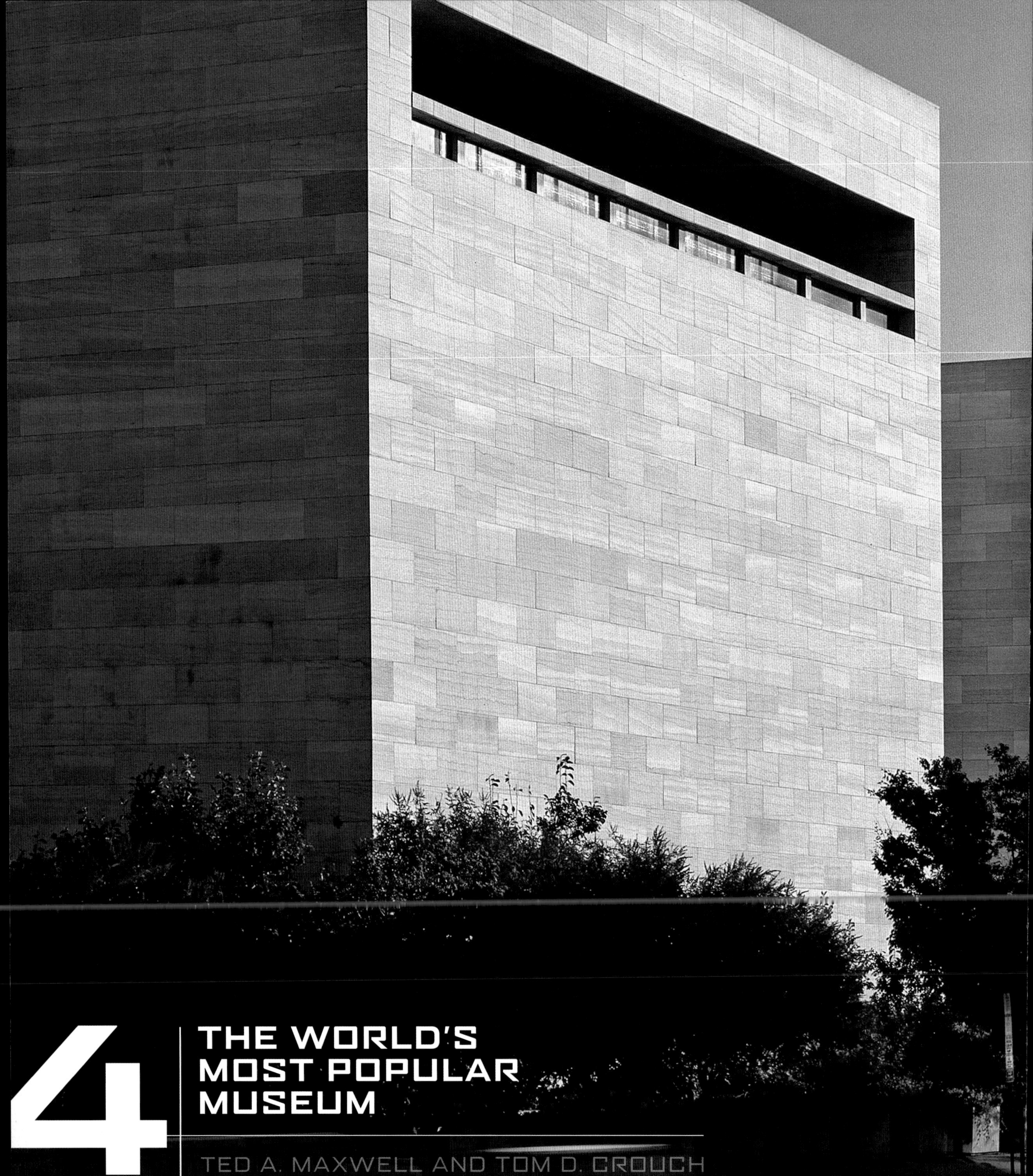

4 THE WORLD'S MOST POPULAR MUSEUM

TED A. MAXWELL AND TOM D. CROUCH

The opening of the museum July 1,

The notion of a Smithsonian air museum had first been discussed in the 1930s. The passage of the National Air Museum Act in 1946 made the dream a reality and opened the door to the possibility of a building dedicated to the history of flight. In the decades that followed, one plan after another was prepared, then discarded. Congress authorized construction in 1966 but did not allocate the $40 million required until 1971. Thirty years of dreaming, perseverance, and hard work came to fruition on that warm and sunny Thursday morning in the Bicentennial summer. / Addressing the crowd gathered outside the north entrance of the building, President Gerald R. Ford pronounced the new museum to be the "perfect birthday present from the

1976, *had been a long time coming.*

American people to themselves," and saluted it as a monument to "the American spirit of adventure." Chief Justice Warren Burger, Smithsonian Secretary Sidney Dillon Ripley, and museum Director Michael Collins offered their thoughts on the occasion, and a signal transmitted by the Viking 1 spacecraft in orbit around Mars reached NASA's Jet Propulsion Laboratory in Pasadena, California, and was sent via telephone lines to the museum. Arriving at the perfect moment, the signal activated a mechanical arm identical to one on the Viking lander that enabled it to gather soil samples. Following a puff of smoke from a heating element that severed the red, white, and blue ribbon, the arm retracted and the National Air and Space Museum was open.

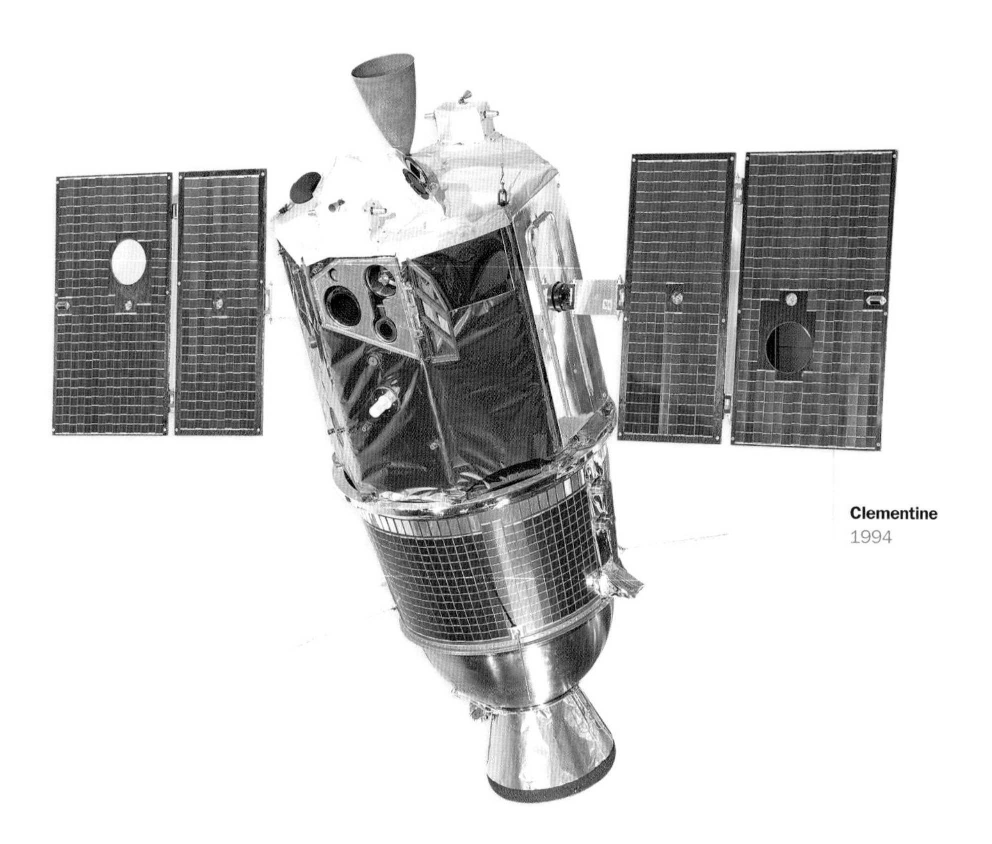

Clementine
1994

FROM THE MUSEUM'S COLLECTION

1970s TO 2000s

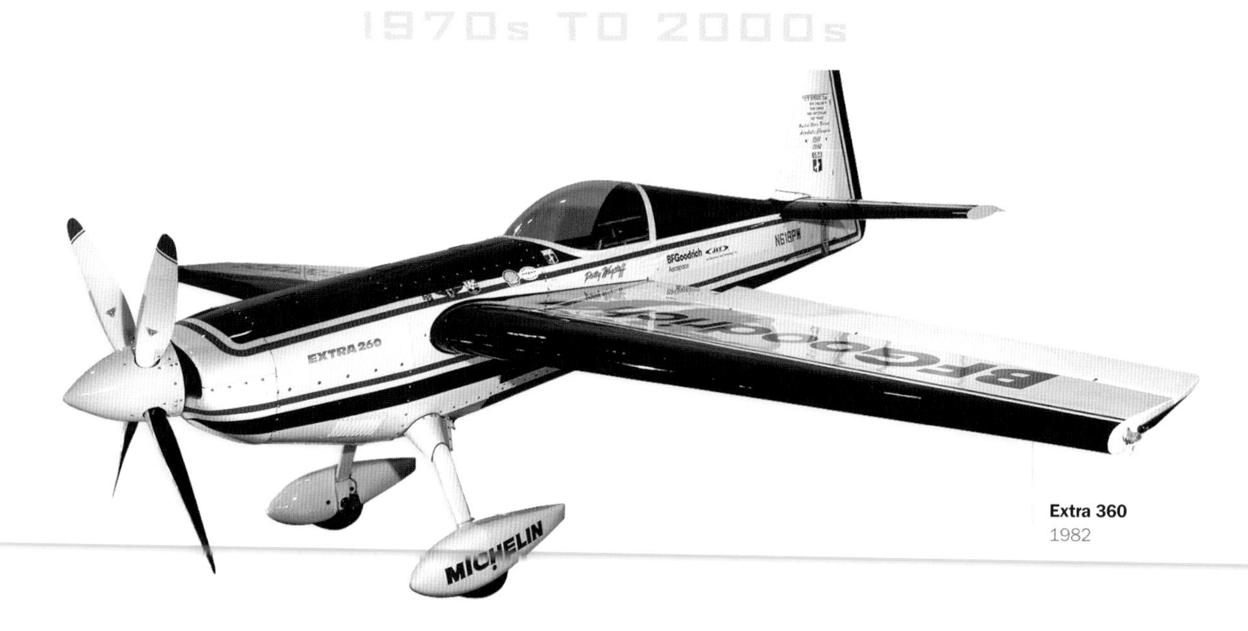

Extra 360
1982

SS-20 Pioneer
1987

Breitling Orbiter 3
1999

MacCready *Gossamer Condor*
1977

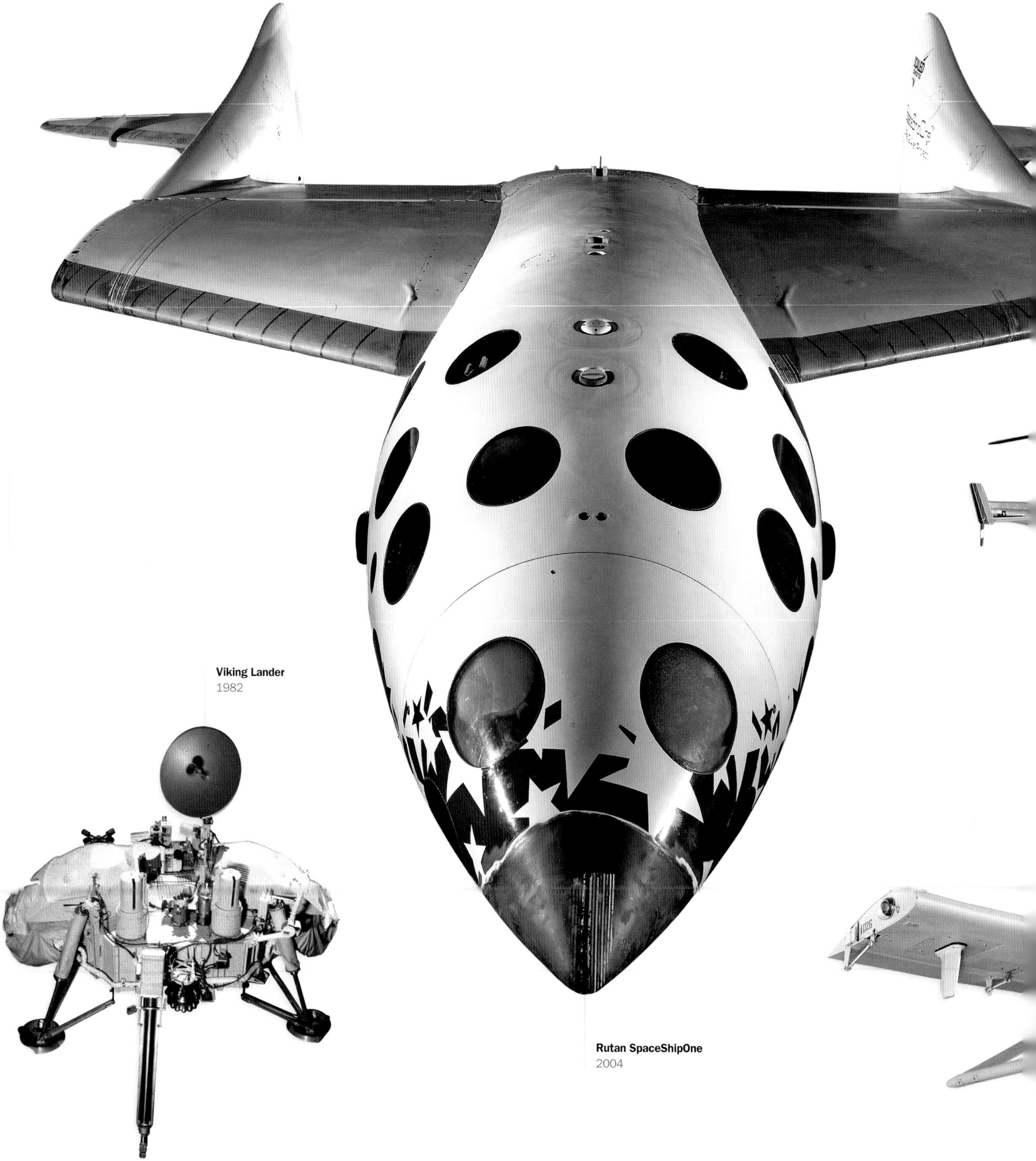

Viking Lander
1982

Rutan SpaceShipOne
2004

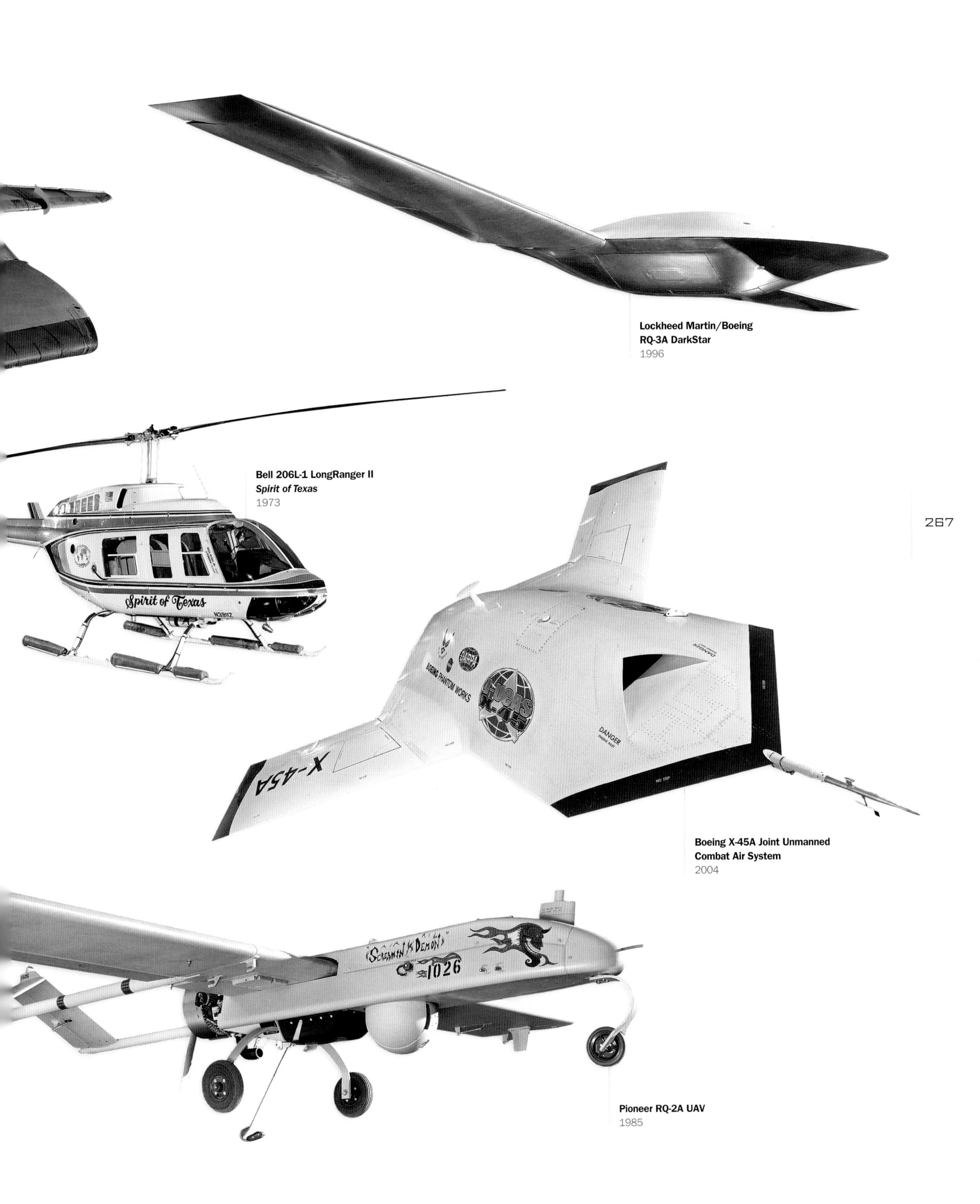

Lockheed Martin/Boeing RQ-3A DarkStar
1996

Bell 206L-1 LongRanger II
Spirit of Texas
1973

Boeing X-45A Joint Unmanned Combat Air System
2004

Pioneer RQ-2A UAV
1985

Architectural critic Wolf von Eckhardt pronounced the new building, with its alternating blocks of pink Tennessee marble and dark glass, "a dignified, handsome work of architecture" and "an appropriate and harmonious finale to the grand architectural concert on the Mall." *New York Times* architectural critic Ada Louise Huxtable offered a more exuberant, but not necessarily more flattering, appraisal. "It's a bird, it's a plane, it's Supermuseum!" she quipped, describing the new National Air and Space Museum building as "a cross between Disney World and the Cabinet of Dr. Caligari."

Walking through the doors that first morning, visitors found ten acres of exhibition space that was anything but dignified and traditional. They entered the building through Milestones of Flight, where they found the icons of the age: the Wright 1903 Flyer, Charles Lindbergh's *Spirit of St. Louis,* Chuck Yeager's Bell X-1, John Glenn's Mercury *Friendship 7* spacecraft, and the Apollo 11 command module that brought Mike Collins, Neil Armstrong, and Edwin "Buzz" Aldrin home from the moon. It was enough to fill the most jaded museum visitor with a sense of awe and wonder.

The element of surprise was never far from the minds of the staff members who had planned the 26 major galleries of the building. "Each gallery is a unit in itself," Mike Collins explained. "I want people to go from one to the next and be surprised by what they'll find. Some are light and frivolous, some are technologically very deep."

Indeed, the displays did not proceed in chronological order. A gallery on the Apollo voyages to the moon could be

With Smithsonian Secretary S. Dillon Ripley (center) looking on, President Gerald R. Ford shakes hands with Chief Justice and Smithsonian Regent Warren Burger at the opening of the National Air and Space Museum on July 1, 1976.

PAGES 260–61: Occupying three city blocks along Independence Avenue in Washington, D.C., the Smithsonian National Air and Space Museum's cubical facade forms a stark contrast to the sleek, aerodynamic artifacts that it contains.

next to a re-creation of a World War I aerodrome somewhere in France. Visitors exited a gallery configured as a cave, in which they could explore the possibility of life elsewhere in the universe, and walked next door to a treatment of exhibition flight and air shows from the early years of the 20th century to the present.

The building had something for everyone. You could touch a real moon rock, walk through a real space station—the three-stories-tall Skylab—or peek through the windows of a replica of the control car of the ill-fated zeppelin *Hindenburg*. Visitors who expected to find displays of historic flight craft were surprised and delighted by the 118 multimedia displays that dotted the museum galleries. "Can we make life?" asked famed television chef Julia Child in one video, as she proceeded to brew up a pot of "primordial soup" using the chemical building blocks from which life may originally have emerged. Automated puppet shows introduced visitors to topics ranging from the technical trade-offs and challenges of aircraft design to the humorous events surrounding the first balloon flight across the English Channel. Ping-Pong balls dropping through a complex vertical maze helped visitors grapple with the complexities involved in transferring Space Age technology to applications that were more down-to-earth.

Collins was not pleased with everything in the new building, however. He was, for example, uneasy about having a deactivated "Little Boy" nuclear weapon of the sort dropped on Hiroshima displayed in a gallery titled Benefits of Flight. "The bomb was part of the museum's collection. We owned it," he remarked to a reporter. "I'm just not comfortable calling it a 'Benefit of Flight.' I think 'Ramifications of Flight' would be better."

OPPOSITE: The backup Skylab Orbital Workshop proved to be an immensely popular walk-through artifact after the opening of the museum. The nose of the V-2 in the background stands in stark contrast to the peaceful use of space.

PROFESSIONAL AIR TRAFFIC CONTROLLERS' T-SHIRT 1981 President Ronald Reagan fired the PATCO strikers for striking illegally.

Many visitors were surprised to discover that art was an important feature of the new museum. They should not have been. "The characteristic art of our time deals with the conquest of space," sculptor Richard Lippold remarked to a *Washington Post* reporter. Lippold's stainless steel sculpture, "Ad Astra," rose high above the museum on the plaza outside the north entrance. Visitors entering the building through the south entrance walked past two great murals painted on the east and west walls of the lobby by Robert McCall and Eric Sloane. In addition, aviation artist Keith Ferris had painted a huge, accurate mural of a single B-17 mission over Germany on the rear wall of the World War II Aviation gallery.

Another Lippold sculpture, "Variations within a Sphere Number Ten: The Sun," a gleaming, 15-foot-tall work, greeted visitors to Flight and the Arts. Paintings hanging in that space chronicled the history of flight as seen through the eyes and experience of some of the great American artists of the century, from traditionalists like Howard Chandler Christy and Norman Rockwell to the decidedly less traditional Robert Rauschenberg. Those whose taste ran to the whimsical could enjoy artist Rowland Emmett's quixotic and wildly imaginative kinetic sculpture "MAUD" (Manually Assisted Universal Deviator)—a whirring, whizzing wonder of a cycle designed, as the label noted, for "moon pedaling by one Professor Leo Capricorn, including zodiac wheel, a pan for sweeping up moon dust, and a cheese sorter."

On the other side of the building, the West German government had donated the Albert Einstein Spacearium as a Bicentennial gift to the nation, complete with a Carl Zeiss Model VI planetarium projector capable of showing 9,000 stars, the Milky Way, and astronomical objects from distant galaxies to the five planets visible to the naked-eye. But it was the IMAX® (Image MAXimum) theater, with its five-story screen and enveloping sound system, that was on everyone's list of must-see experiences. The premier film, *To Fly!* produced by James Freeman and Greg MacGillivray, carried visitors on a spectacular journey that began with a balloon flight and ended in space. When museum officials tried to remove *To Fly!* from the regular schedule in the 1990s, they received such a barrage of complaints that they chose instead to upgrade the original film with digital sound and projection, rejuvenating the first great IMAX classic.

To say that the museum was an instant success is to understate the matter. As a *New York Times* reporter commented, "The only thing missing is a feather from Icarus's wings." On July 4, three days after the opening, 65,000 people walked through the doors. Attendance totaled 87,000 one day during that first month. Five times that day, security officers had to turn people away when the building reached its internal capacity. The five millionth visitor, Theresa Davis, of Hyattsville, Maryland, arrived at 1:30 p.m. on December 30, 1976, just six months after the opening. Assistant Secretary of the Smithsonian David Challinor and Deputy Director | *to page 274* |

The new National Air and Space Museum became the most popular museum in Washington and the world. Within 25 days, more than a million people had come through the doors, a number that has increased to more than a quarter billion in the more than three decades since opening.

| *from page 270* | Mel Zisfein were waiting to greet her. "The Friday after Thanksgiving was like gangbusters in here," Zisfein remarked to reporters. "And to think that before this place was built some people in Washington asked: 'Who's going to come to see a bunch of old airplanes?' " By the end of the first year, museum attendance had reached 9.6 million, an average of 26,000 visitors a day.

Shortly after the museum opened, Michael Collins sent a memo to the staff in which he wrote, "I thought I would share with you some of the comments I have read regarding our new museum. Thank you for making them possible." Among others, he quoted Barry Goldwater, who wrote effusively: "Today is one of the greatest days in the history of aviation . . . together under one roof, evidence that man, using courage and thoughtfulness along with God's guidance can achieve anything." S. Dillon Ripley remarked, "The opening of the NASM has been the culmination of many years of planning and cooperation among many people. . . . The ceremonies connected with this great event were just right. . . . My deep appreciation for making it possible."

The national press was equally enthusiastic. *Newsweek* magazine called it "a building that has drawn critical acclaim. Even more impressive," the writer said, is that under the watchful eye of Michael Collins and Mel Zisfein, "the museum was completed ahead of time and within its budget." The *New York Times* wrote that "Washington and the Smithsonian have finally moved into the 20th Century. . . . The exhibits defy description," and include "every audiovisual and electromechanical device known to man." The newspaper predicted that "the exhibits will bring joy, instruction, and wonder, to children of all ages."

OPPOSITE, TOP: NASM Deputy Director Don Lopez gives a tour of the museum to singer Michael Jackson in 1984. Visits by stars and heads of state are common, keeping the security force busy.

OPPOSITE, BOTTOM: When spacesuits were first displayed in 1976, curators did not know the issues of longevity and the brittle failure of the plastic components associated with them. Now they are rotated and displayed with helmets and gloves off.

MUSEUM TROPHY: "THE WEB OF SPACE" 1985
One of the trophies annually presented by the museum for current achievement and lifetime achievement in aviation and space.

The leaders of the aerospace industry and its press representatives could hardly contain their joy. The president of the Grumman Corporation wrote, "While I had read newspaper accounts . . . , I was unprepared for the impressive scale and the superb attention to detail in arranging the exhibits." *Aerospace* magazine added, "It is appropriate that America's birthday gift to itself is a magnificent new home for the . . . National Air and Space Museum."

The extent of popular enthusiasm for the new museum was obvious, but museum leaders were determined to discover precisely who was coming, and exactly what they liked and disliked about it. Collins and Zisfein commissioned Yankelovich, Skelly and White, a Madison Avenue marketing and research firm, to do "A Study of Visitors' Reactions to the National Air and Space Museum and the Individual Exhibits."

The study, begun in December 1976, was done in four phases and consisted of 4,000 interviews with visitors to the museum. The results indicated that the museum was rated "an outstanding success" (with 98 percent saying it was as good as or better than expected). Visitors judged the museum to be "meaningful experience," and said it was "more educational, more interesting, more fun, more colorful than they had anticipated." Initial information indicated that the museum appealed to visitors that had an interest in the sciences, space, history, and travel, and that more than half of the visitors surveyed had made a repeat visit to the museum. Visitors were also impressed with the museum's appearance, size, and spacious corridors; the physical setup of the exhibitions; the helpfulness of the attendants, explanations, and descriptions; the names of the exhibitions; the heating and ventilation; and the ease of finding what they wanted in the museum. Less impressive were the restrooms, gift shop, and cafeteria.

MOVING FORWARD

Perhaps more significant than the initial visitor reactions to the museum were the behind-the-scenes plans that Collins, Zisfein, and the staff created to replace and upgrade exhibits and to develop a program of research at the museum. The museum organized a Visiting Committee, made up of such notables as Gerard O'Neill of MIT, retired astronaut Russell L. Schweickart, and Richard Whitcomb of NASA's Langley Research Center, to advise on the programs, policies, and research to be carried out at the museum.

The research program called for investigation of a number of subjects, including the history of NASA's Dryden Flight Research Center; the history of McCook Field, an early center of aeronautical research and development; the subject of technology transfer from flight industries to the general economy; the role of women in aviation; the history of the first African-American fighter squadron in the USAAF; and plans to develop a source guide for the study of aeronautical history.

The establishment of a program of academic fellowships during the years immediately after the opening was a critically important step toward building the scholarly reputation of the

National Air and Space Museum. The creation of Guggenheim Fellowships, one-year grants intended to promote scholarly research in aerospace history and technology, attracted the best talent in the field to the museum. Professor William M. Leary, Jr., received the first competitive Guggenheim Fellowship in Aerospace History to support his work on the history of commercial aviation in the United States between the World Wars. The next year, the museum awarded two Guggenheim Fellowships. One went to Von Hardesty, a specialist in Russian history and a professor at Bluffton College who would later join the museum's curatorial staff. The other fellowship went to Valnora Leister of São Paulo, Brazil, thus beginning a tradition of inviting individuals from around the world to participate in the museum's "company of scholars."

Several other early museum fellows were destined for great things. Joseph Corn shaped his classic book, *Winged Gospel*, during his year at the museum, while Walter McDougall used his time as a fellow to hone the ideas that would form the backbone of his Pulitzer Prize–winning history of the early Space Age, *The Heavens and the Earth*. Thirty years later, the combination of Smithsonian pre- and postdoctoral fellowships and the Guggenheim Fellowship program continue to draw fresh blood and new ideas to the museum.

Not content with attracting young scholars to his museum, Mike Collins instituted a rotating Charles A. Lindbergh Chair in 1977 as a means of luring the most senior scholars, engineers, and scientists to the museum for a year. He could not have made a better choice to inaugurate that tradition than Charles Harvard Gibbs-Smith. Retired from a distinguished career at London's Science Museum, Gibbs-Smith was a polymath with a special taste for the early history of flight. His masterwork, *The Aeroplane: An Historical Survey*, remains today the standard history of the subject. His year at the museum was an important step in establishing the historical credentials of the museum.

To a great degree, the museum was able to maintain the high standards of the Lindbergh Chair. Benjamin S. Kelsey followed Gibbs-Smith. One of the most distinguished engineer-pilots in the history of the U.S. air arm, Kelsey was an MIT graduate who worked with Jimmy Doolittle on the early blind flying experiments; shepherded the P-39, P-38, and P-51 into existence; served as chief engineering officer of the wartime Eighth Air Force; and played a key role in the Air Force's postwar planning. He summed up a lifetime of experience and lessons learned in *The Dragon's Teeth?* published posthumously in 1982.

Other distinguished Lindbergh Chair holders include Hans von Ohain, the German inventor of the turbojet engine; John Fozard, the English engineer who pioneered vertical takeoff and landing aircraft; Pierre Lissarrague, historian of French military aviation; and W. David Lewis, the distinguished American historian of aviation. After three decades of struggling to find and build a permanent home for its collections, create significant exhibitions, and develop a plan of research and scholarship, the National Air and Space Museum was on its way to achieving the success and respectability that had eluded it for such a long time.

There was little time to rest on post-1976 laurels, however, | *to page 280* |

The U.S. Navy's Blue Angels display their skill in piloting Douglas A-4F Skyhawks in this still frame from the IMAX film *To Fly!* (1976), the museum's first and most popular film, which chronicles flight from ballooning to the Space Age.

Don Lopez was a World War II ace, having shot down five Japanese aircraft in China while flying the P-40 and P-51, in the process earning a Distinguished Flying Cross. After the war, he was a test pilot, served a short combat tour in Korea flying F-86 jets, received a bachelor's and a master's degree in aerospace engineering, and worked on the Apollo and Skylab programs.

DONALD S. LOPEZ

ON THE LINDBERGHS

He joined the museum in 1972 as the head of the Aeronautics Department and became deputy director for four different directors (Walter Boyne, Martin Harwit, and, after a hiatus, Donald Engen and John Dailey); he was acting director in 1999–2000. He served the museum faithfully until his final flight in 2008.

The following excerpts are taken from his books, *Into the Teeth of the Tiger* and *Fighter Pilot's Heaven.*

. . .

My earliest memory is of an event that sparked both a lifelong passion and my profession. At about three and one half years of age, I was taken by my parents to a major highway in Brooklyn. After sitting on the curb for a long time, in a big crowd, I was caught by a surge of excitement, stood up, and waved at a man in an open car to the accompaniment of loud cheers of "Lindy, Lindy!"

Time has erased other details of that incident, but the image of that handsome, heroic figure has stayed with me. Somehow I must have learned that he was a flier. I cannot remember a time since when I was not interested in flight.

I was fortunate in that my growing-up years coincided with the golden age of flight, that period when almost everything that took place in the sky was news. Speed records were set, races were won, oceans were crossed, and the pilots who flew those magical machines were heroes to youths and adults alike. . . .

At times, things in life seem to go full circle. . . . On the day of my retirement from the Air and Space Museum I was honored to escort Anne Morrow Lindbergh on a brief tour of the museum. [Lopez never actually left NASM, but kept an office, and later "unretired" to become Engen's deputy.] *Accompanied by her daughter Reeve, she arrived early, and I went with them to the second floor. There she paused at the* Spirit of St. Louis *and the exhibit of Lindbergh memorabilia. She then went to a bench opposite the Lockheed Sirius* Tingmissartoq, *in which she and her husband had spent so many hours exploring new air routes to Asia and Europe.*

I stood a few feet away and watched as she sat, serene and lovely, with tear-filled eyes, quietly reliving some of their adventures together. I, of course, did not interrupt her reverie, but I hope that she realizes how universally she is loved and admired for her courage, devotion, and wonderful writing. It was truly a privilege to be with her. To me, she is unquestionably the woman of the century.

| *from page 276* | for there is no such thing as a static air and space museum. New records were being set, new prizes won, and the deeper realm of space remained to be explored. Only a year after opening, the Paul MacCready–designed *Gossamer Condor* became the first human-powered aircraft to navigate the prescribed 1.3-mile course, winning the Kremer Prize and a spot in the museum. However, the fragile, 70-pound aircraft composed of aluminum tubes, Mylar, and guy wires was never designed for an air-conditioned building that hosts nine million people a year. Over the years, blasts of air, the opening of the bay doors for artifacts and equipment, and the dust of millions took their toll on the thin plastic.

When the museum's windows were replaced in 1998, the aircraft was restored at the Garber Facility before being moved to a calmer area of atmospheric circulation at the museum. MacCready's *Condor* was later united in the collection with his *Gossamer Albatross,* the first human-powered aircraft to cross the English Channel, and the *Solar Challenger,* a 200-pound aircraft that made the first solar-electric powered flight from France to England in 1981.

Aeronautics was not the only subject for records. With the Viking orbiters circling Mars and the two Viking landers firmly on the surface of the planet since shortly after the museum opened, plans were under way for an ambitious mission to the outer solar system. Two Voyager spacecraft were launched in 1977 to fly by Jupiter and Saturn, and the test spacecraft was acquired from the Jet Propulsion Laboratory to highlight a new Exploring the Planets gallery. What wasn't known then was the number of changes that gallery would need as a

Human-powered flight had been a dream even before Leonardo da Vinci's sketches of flying vehicles but was not realized officially until 1977, when the *Gossamer Condor* became the first to fly a prescribed course and win the Kremer Prize for human flight.

result of the discoveries from that program. After flying by Jupiter and Saturn, Voyager 1 was directed out of the ecliptic to explore interstellar space, but not before discovering active volcanoes on Io, lightning and massive storms on Jupiter, new rings of Saturn, and moons that defied interpretation. Voyager 2 went on to fly by Uranus in 1986 and Neptune in 1989. Both spacecraft continue to return data on the interactions of our sun and interstellar space, and they may not lose contact with Earth until 2020.

The new aerospace triumphs meant growing collections and new challenges for the museum. Whereas some people assumed that the new building would house all of the collections and solve the poor storage issues, the staff knew that less than a quarter of the aircraft and only a small percentage of other artifacts would fit in the new building. The rest would remain in obscurity in Silver Hill, Maryland.

In 1975, Collins had given the responsibility for the collections' move to retired Air Force Col. Walter Boyne, then a curator in Aeronautics. Boyne had an intimate knowledge of the collections, and he and the restoration staff at Silver Hill had the formidable job of moving more than 60 aircraft, spacecraft, and missiles. That this was accomplished within six months, aided by engineers from nearby Fort Belvoir and a contract crew from United Rigging, was a minor miracle. But what happened next was even more astounding. Where once weeds and small trees grew through gaps in aircraft stored outside for decades at Silver Hill, a new museum was created.

DUPLICATE VOYAGER RECORD COVER, "SOUNDS OF EARTH" 1977 Mounted to the exterior of the Voyager spacecraft as a message to alien civilizations

OPPOSITE: First color image from the surface of Mars taken by the Viking 1 lander. A hint of layering in the distant rocks was confirmed by later spacecraft, establishing that the planet has had a varied history of water and lava deposits as well as impact cratering.

By 1977, the restoration staff at Silver Hill had moved nearly all the aircraft to inside storage (a notable exception was John Kennedy's campaign plane, the Convair 240 *Caroline*) and opened two buildings for guided tours. In the coming decade, 17 more buildings were made available to the public, and three-hour docent tours became the norm. While most aviation and space enthusiasts loved the tours, some accompanying family members returned to the waiting area to escape an extended conversation about landing gear and ball turrets that would try the patience of even the hardiest fan. With several hundred a week taking these tours, the museum decided on an additional option—the Open House. Until preparations for the Udvar-Hazy Center forced their cancellation, one weekend a year in late April would see at least 10,000 visitors tour 19 buildings of satellites, kites, wings, fuselages, engines, and all manner of technological history.

Even before the museum opened, the restoration of artifacts that passed through the hands of the museum craftspeople at Silver Hill was becoming professionalized. In March 1973, Mike Collins invited leading aerospace engineers, museum officials, and corporate specialists to a restoration planning meeting. The attendees were asked to respond to some basic questions: Should aircraft in the museum collection be flown, or restored to flying condition? To what extent should original parts or structures be replaced with new material? Should the markings in which an actual aircraft flew be replaced by those of a more historic example of the same type? What should be the policy on engine restoration, and the preparation of cutaway displays of engines? Should old or damaged fabric be replaced?

The answers to those and other questions, developed as a result of extended discussions among those attending the meeting, continue to provide basic guidance to museum curators and craftspeople to this day. The approach to the treatment of historic aircraft has continued to evolve, however. In the years before and after the opening of the museum, the most common approach was a full-blown restoration, an attempt to return an aircraft to the pristine appearance it had when first in use. All too often, the result was an airplane that looked better than it did on the day it rolled out of the factory. Over the next 30 years, however, the practice of treating historic aircraft has evolved further. Before undergoing treatment, each artifact is assessed independently and a precise treatment plan is developed jointly by the curator, conservator, and restoration technician. Preservation or conservation is usually preferred over full-blown restoration, as evidenced by recent projects such as the museum's World War I–era Curtiss JN-4 "Jenny" and Caudron G.4 French two-engine bomber, which have been as far as possible preserved in their original condition.

The late 1970s and early '80s proved to be a time of transition for the museum administration too. After Mike Collins moved to the Castle as undersecretary of the Smithsonian in 1978, the next director, Noel W. Hinners, was appointed. Hinners came from NASA, where he had been associate administrator for space science. Arriving in 1979, he promoted historical research,

instituted the first peer-review system for museum scholars and, with Walt Boyne as executive officer, pushed through a reorganization that folded Science and Technology into the Aeronautics and Astronautics Departments (the latter renamed Space Science and Exploration, and eventually, Space History). Mel Zisfein and Fred Durant were sidelined and soon retired. But Hinners's heart was in space missions, so when the directorship of Goddard Space Flight Center opened up three years later, he moved back to NASA.

Walt Boyne was named acting director of the museum in 1982, and he succeeded Noel Hinners as director the following year. Secretary S. Dillon Ripley cautioned Boyne, a military pilot, on the need for a new emphasis on the research that would provide the foundation for future collecting and exhibitions. Boyne promoted a curator with strong scholarly credentials to oversee the museum's research and publication programs and initiated an effort to produce a multivolume history of aviation. While the project ultimately fell short of the goal, it did lay the foundation for the creation of a series of scholarly studies in aviation history in 1989. The series, a joint venture between the museum and the Smithsonian Institution Press, attracted the very best manuscripts to the field and marked the museum as an international center for serious research and publications on the history of flight. In an effort to reach beyond the walls of the museum, Boyne also launched *Air & Space/Smithsonian* magazine, which published its first issue in 1986. One of the most successful forays into magazine publishing of the era, *Air & Space* continues to inform and entertain readers around the world on topics directly related to the interests of the museum, although it is now published by Smithsonian Enterprises and not the museum.

Boyne was, moreover, a strong advocate for the archival collections and for the use of new technology. With over a million photos in the collection, and the latest in (analog) technology in the form of the 12-inch videodisc, the newly formed Records Management Division spent countless hours rephotographing each picture and mastering videodiscs for the museum's as well as other organizations' use. The four discs produced could be accessed using "step" motion on the videodisc player; later, a computer interface was added for local use.

In time, the Records Management Division metamorphosed into a full-fledged Archives Division, with staff trained in archival processing and care. The mergers, acquisitions, and failures of aircraft manufacturers and the reorganizations of federal agencies swelled the volume of material donated to the museum. Even with much of it in the form of microfiche and microfilm, acid-free boxes now lined the shelves from floor to ceiling of an entire building at Silver Hill, containing everything from photos, plans, film clips, blueprints, millions of technical drawings, and 1,600 cubic feet of technical manuals.

Today, the Archives Division is much more than a static repository. Coming from a long tradition of service to the community and general public, the staff answers thousands of inquiries each year, ranging from paint colors to screw sizes. In both public inquiries and the ongoing work of categorizing

OPPOSITE: By 1998, free space for storage was becoming hard to find at the Paul E. Garber Facility. Wings were stored on racks in buildings separate from fuselages to save space, and crates were stacked in all open areas.

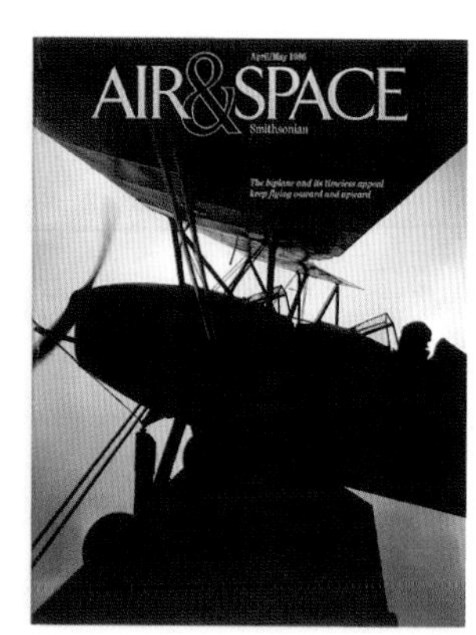

***AIR & SPACE/ SMITHSONIAN* MAGAZINE APRIL/MAY 1986**
Front cover of the first issue

and housing the collections, it rapidly became apparent that there were no standards or cross references for aircraft designations. Dealing with 34,000 types of aircraft produced by 6,600 companies around the world forced the staff to create the *Directory of Airplanes,* a cross-index of aircraft models and manufacturers that is now the standard for all aviation archives. But the restoration staff members always have their noses in the archives, and with each project, the original blueprints and technical files are used, if possible.

In the early years, the museum kept a fast pace, offering new exhibit galleries (nearly one per year), new IMAX films, and other public symposia and programs. It was easy; the revenue from the popular IMAX films as well as from the shops all came directly to the museum, which had the autonomy to run the shops and reinvest in new films. The excess proceeds went to new exhibits and planetarium shows, and sizable amounts were added to the museum's fellowship endowments—investments that would pay off for years to come.

By the mid-1980s the need for a new offering in the large-screen format became obvious. The two films the museum had made since *To Fly!* were—to put it charitably—lessons in how not to make a large-format film. *Flyers,* with a contrived plot that distracted from some excellent stunt flying, and *Living Planet,* with superb footage that couldn't triumph over a too-familiar message, simply did not measure up to the premiere film.

Thus, when IMAX producers Graeme Ferguson and Toni Myers approached NASA and the museum

1980-423
1948-178(2)
1980-423

FA00153
FLIGHT INTO
FA 01282 TWO SCENES
ARMED PORTER
FA 02420 INDUSTRIAL

In addition to millions of photos and documents, the NASM Archives contains films of early aviation and spaceflight, often sought after by commercial producers. The film is kept at a controlled low temperature to reduce deterioration.

about the possibility of flying an IMAX camera on the shuttle, there was plenty of interest. The devilish details of allowing a proprietary Canadian camera on board the U.S. space shuttle, where all photographs had heretofore been in the public domain, and producing a film with a NASA contractor as a corporate sponsor, literally took years to negotiate. The results were worth it. *The Dream Is Alive,* when first viewed by shuttle astronauts, was regarded as the "next best thing to being there." It was followed by *Blue Planet* and *Destiny in Space,* and while not on the manifest for every flight, the IMAX camera has since filmed such events as the deployment of the Hubble Space Telescope and its several repair and refurbishment missions.

In 1985, at the suggestion of relatives of the Wright brothers, the museum staff disassembled the world's first airplane, the 1903 Wright Flyer, and conducted the most extensive study of the historic machine ever undertaken—all in full view of the public. The result was a wealth of new information regarding the centerpiece of the collection, an opportunity to document the condition of the craft (and the assurance that it was in good condition), and the return of the machine to its appearance at the time of its flight.

Several of the original 1976 exhibitions proved less than fully successful, and so they were replaced by second-generation exhibits incorporating the lessons learned in the run-up to opening. The new displays included Exploring the Planets, Early Flight, Jet Aviation, and General Aviation. Beyond the Limits: Flight Enters the Computer Age, which opened in 1989, marked the introduction of new media into museum exhibitions. No longer content to place objects on the floor with static labels, this gallery sought to provide visitors with an immersion experience. The computer-dependent NASA experimental X-29, with its unique forward-swept wings, hung from the ceiling in Beyond the Limits, where visitors encountered computer-based interactives ranging from aircraft design to space mission trajectories.

NEW CHALLENGES

The second decade of the mall museum would see many changes—in management, exhibits philosophy, and outreach. Following the departure of Walt Boyne in 1986, James Tyler, who had previously served as acting director of the National Museum of Natural History, agreed to serve in the same role for the museum while Smithsonian Secretary Robert McCormick Adams searched for a new director.

In 1987, Adams appointed Martin Harwit, a Cornell University astrophysicist and former occupant of the museum's short-lived Martin-Marietta Chair in Space History, to the museum post. The new director established staff committees to assist in planning for the future of the museum. He would also wrestle with changes imposed by Smithsonian leadership. Management of the museum shop was transferred to a central organization, and with it went some of the income that had once supported new exhibits and fellowships. Now that the museum was increasingly forced to rely on donations to build new exhibitions and buildings, the director's development role grew. Harwit created a Development Office to fill the funding gap created by the loss of traditional income.

The new team inherited the need for an expansion in collection storage and display space. During the Boyne years, planning had begun for a new facility to be based at a major airport, the subject of the next chapter. The arrival of the space shuttle *Enterprise* at Dulles International Airport in 1985, and the Lockheed SR-71 reconnaissance aircraft, which landed there following a record transcontinental flight in 1990, helped cement that location as the site of the center. Meanwhile, the downtown museum acquired several major artifacts during the Harwit years. In December 1987 Burt Rutan's *Voyager* aircraft, which his brother Dick Rutan and Jeana Yeager flew around the world nonstop without refueling, was hung in the south lobby entrance a year after its record-setting flight. In May 1990 the museum installed the Soviet SS-20 and American Pershing II intermediate-range missiles in Milestones of Flight. They marked the end of the Cold War and the first treaty that eliminated a whole class of nuclear weapons.

Harwit also emphasized the need for a new approach to exhibitions that would both entertain visitors and help them understand critical issues. The museum's art department, now renamed the Department of Art and Culture, led the way with a 1992 exhibition that explored the way in which the 1960s television program *Star*

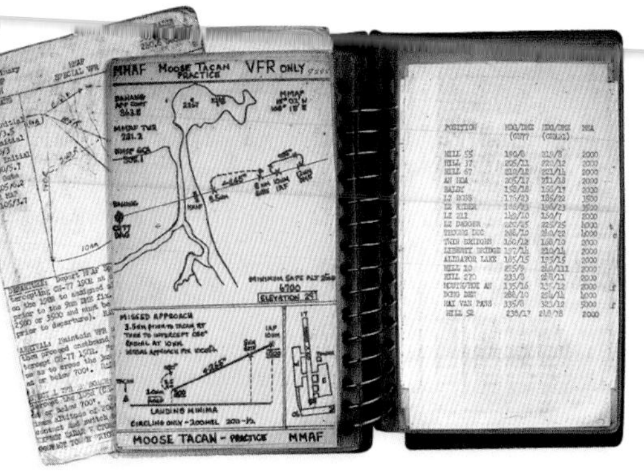

CREW CHECKLIST NOTEBOOK Ca 1970
Used by Sikorsky CH-53A helicopter pilot Robert S. Vieta during his tour of duty in Vietnam

TOP: The Paul E. Garber Facility was certainly a "no frills" museum in the late 1970s and 1980s. Docents led tours of the unrestored artifacts and aviation oddities.

BOTTOM: Each year the collections staff is aided by interns who learn the skills and techniques of preserving and restoring aviation and space artifacts. Here, interns are applying stencil to the Boeing B-29 Superfortress *Enola Gay* in 2001.

Trek addressed controversial social and political issues, all within the relatively safe context of an entertaining science-fiction format. A follow-on exhibition five years later described the way in which the colorful costumes and characters of George Lucas's *Star Wars* film series related to classic mythological archetypes. Their serious messages notwithstanding, the two exhibitions set new records for attendance at the world's most visited museum.

Another offering from that department proved more controversial. Flight Time Barbie, opening in 1995, consisted of two cases displaying Barbie and Ken in the surprising number of flight-related costumes in which they had been marketed over the years. Andrew Pekarik, of the Smithsonian Institutional Studies Office, was fascinated by the divergent visitor reactions to the display. "The diversity of opinions on Flight Time Barbie is striking, [ranging] from adoring to hostile. . . . In this public setting, elevated through its presentation at the Smithsonian, Barbie became a focal point for issues of gender-identity, self-image, the power of popular culture, and the nature of the National Air and Space Museum."

The approach of the 500th anniversary of Christopher Columbus landing in the New World inspired museum curators to look to the future of space exploration rather than the past. By concentrating on the possibility of a future mission to Mars, Where Next, Columbus? posed a key question: "Is it better to explore with robots, or with people?" Visitors came to appreciate the difficulties of long-duration space travel, from radiation to food, and included a garden—a hydroponic assortment of lettuce and greenery grown totally

The restoration shop at the Garber Facility—here shown some years before the move to Udvar-Hazy, in 2010—was the place where the collections staff always had a variety of wood and metal projects under way, utilizing their expertise in different areas of restoration. This day, work was ongoing on (in the foreground) the Hawker Hurricane and the Nieuport 28.

in artificial light. Designed for a two-year stay in the museum, Where Next remained in place for ten years, until it was removed to make way for another anniversary exhibition.

The process of crafting exhibitions has certainly changed over the years. In 1976, a curator and a designer worked together to create exhibitions. Today professional educators have joined them as key members of exhibition development teams. The museum's Education Department played an especially important role in the conception and operation of a new How Things Fly gallery, which opened in fall 1996. Filled with imaginative interactives that help youngsters of all ages grasp the basic principles of flight in air and space, How Things Fly is also the home of a new group of museum staffers, the "explainers," a corps of young people who interact directly with visitors. An amphitheater was built into the space, presenting entertaining talks on scientific principles as well as paper airplane contests and other live events designed to entertain and instruct. All in all, a visit to How Things Fly is the sort of hands-on experience undreamed of by those planning the opening exhibitions in 1976.

THE WORLD'S MOST UNPOPULAR MUSEUM

Despite the public acclaim and record attendance that marked the early years of the museum, there were persistent critical voices. Three years after the opening of the new museum building, Howard Learner, in a report to the Center for Science in the Public Interest, took a dozen-odd museums of science and industry to task for their failure to put scientific and technical achievement in a responsible social, economic, and political context. The National Air and Space Museum, he charged, was little more than "a temple to the glories of aviation and the inventiveness of the aerospace industry."

Historian Michal McMahon agreed. "The promotion of technological progress and the resort to 'hype' underlies most of the Air and Space Museum," he explained to the readers of a leading academic journal in 1981. He noted the "omission of the 'offensive' themes in twentieth-century history represented by the World War II bomber," and suggested that the museum "should be concerned with both the positive and negative aspects of technology." In 1990, historian Samuel Batzli pointed to the admonition of the Smithsonian advisory council that the Institution as a whole should present a "critical and non-celebratory attitude towards science and technology."

Indeed, museum staffers had struggled to achieve a balance between the competing roles of their institution as a shrine in which to celebrate past achievements and a classroom in which visitors can be encouraged to recognize the true nature of the difficult process of technical innovation and the complex impact of technology on society. They took one small but significant step in 1982, with the opening of the pathbreaking Black Wings exhibit and accompanying book by Von Hardesty and Dominick Pisano. The exhibition explored the long struggle of black Americans to take to the sky, from the Jim Crow years of the 1920s to the heroic era of the Tuskegee airmen of World War II. Black Wings marked a new willingness on the part of museum administrators and staff to address difficult social issues.

Continued attempts to address issues of balance seldom drew public attention. But in 1990, the Space History and Exhibits divisions crafted a new mini-exhibition for the museum's V-2 missile, noting that the rocket that had opened the way to space had been designed as a Nazi terror weapon and built by slave laborers, thousands of whom died in the process. An accompanying photo showed victims of the weapon lying dead on an Antwerp plaza. The *Washington Post* heralded the change as "a rare breakthrough" in "truth in labeling" for "a temple of aerospace lore where tradition has called for bland captions on horrifying instruments of war."

A year later, Aeronautics curators took an even larger step, replacing an older gallery on World War I aviation. The original exhibit was one of a few in the museum that attempted to be immersive, but the three airplanes, mock hangar, and background mural were not up to contemporary exhibit standards, nor did the content meet historical ones, based as it was on the popular mythology of "knights of the air" divorced from ground combat. Even worse, during a routine artifact inspection, moths were found infesting a wool uniform, and it was feared they might have spread to the wood plank floor and other parts of the gallery. The exhibit was dismantled immediately, and the barren gallery was used to show off several World War I aircraft until a new exhibit could be completed on the other side of the building.

Legend, Memory and the Great War in the Air opened in late 1991. A labyrinth of narrow alleys gave the feel of trench | to page 298 |

During its restoration on the floor of the museum in 1985, details of the original construction of the 1903 Wright Flyer were recorded to document the Wright brothers' techniques, as shown in this contact sheet. Several parts of the airplane had been fixed over the years after its historic flight.

MAGELLAN T. BEAR
1995
Flew on the space shuttle *Discovery* STS-63 mission in February 1965

The Rutan Model 76 *Voyager* on display in the lobby of the National Air and Space Museum. In 1986 the airplane made the first nonstop, non-refueled flight around the world.

On December 23, 1986, pilots Dick Rutan and Jeana Yeager completed the first nonstop, non-refueled flight around the world in *Voyager*, an aircraft constructed almost entirely of composite materials. It had taken two years to convert designer Burt Rutan's napkin drawing of a flying fuel tank into an airplane, and two more to plan the flight.

THE RUTAN VOYAGER

FIRST NON-REFUELED GLOBAL AIRCRAFT

In cramped and uncomfortable conditions, the pilots endured nine days of slow and stressful flying. But was it really one of the last great records of aviation? Winning the prestigious Collier Trophy seemed to settle that discussion, and the museum decided to collect this craft.

In the summer of 1987, *Voyager* was dismantled for its trailer trip from California to the Garber Facility. The aircraft received accolades during its showing at the Experimental Aircraft Association convention in Oshkosh, Wisconsin, but structural engineer and NASM curator Howard Wolko still had to calculate how to get this huge aircraft into the museum building.

After a midnight wide-load transport from Silver Hill to the museum's

west patio, all the carefully laid plans went awry. Just inside the doors, a replica aircraft carrier deck exhibited in the museum's Grumman Hellcat protruded a little too far, and *Voyager's* center section, wheeled on dollies, could not get past.

In the wee hours of the morning, restoration specialists found a solution: Elevate and tilt the section with a hydraulic lift, inching it over and past the carrier deck. After barely sliding it by the Air Transportation gallery, they rolled it into the South Lobby at dawn. Fortunately, assembling the wings, empennage, and engines was routine. The exhausted staff suspended *Voyager* using scissors lifts and winches in time for the ten o'clock opening. *Voyager's* near-catastrophic loss of the winglets on takeoff proved fortunate, as it reduced the wingspan by two feet, allowing the aircraft to fit snugly under the roof of the lobby.

Pilots Dick Rutan and Jeana Yeager flew around the world in nine days, enduring very cramped quarters and harsh conditions. They received the Collier Trophy for one of the last great aviation triumphs.

| *from page 292* | warfare, while above, a restored Fokker D V.II, an SPAD XIII, a Sopwith Snipe, and a Voisin VIII, among others, were a stark counterpoint. The gallery underscored the links between the war on the ground and the one in the air; focused on the need to enlist the total economic and industrial power of a nation to fight an air war; suggested ways in which the lessons of World War I played out over the remainder of the 20th century; and explored the extent to which popular culture, nationalism, and even the requirements of postwar commerce shaped a collective memory of World War I aviation. Instead of romanticizing fighter pilots, the gallery paid equal weight to the failures and successes of airpower, and ended by discussing the legacy of strategic bombing, which had already begun in 1915.

The ability to produce an important exhibition on aerial warfare that went beyond a celebration of valor or a discussion of technological mastery was heady stuff, and set the stage for the most explosive controversy in the history of the museum—perhaps in the history of American museums. In a critical assessment of the museum published a decade before, historian Michal McMahon had suggested that the museum could begin to move "into a position that respects history and hence respects its visitors," by responding to a simple question: "Why not the *Enola Gay*?"

Indeed, the B-29 that dropped the atomic bomb on Hiroshima was one of the best known and most controversial airplanes in the national collection. In 1970, while spearheading the fight to obtain congressional authorization for a new museum building, Senator Barry M. Goldwater expressed a vision of it as "a patriot's museum," a shrine celebrating aerospace achievement. Surely, he commented, the *Enola Gay* was not the sort of "truly historic" aircraft worthy of display in a collection that would "serve as both a remembrance of past accomplishments and an inspiration to our young citizens that the age of discovery is not over."

OPPOSITE, TOP: The space shuttle *Enterprise* perches on the ferry 747 upon arrival at Dulles International Airport in 1985. The *Enterprise* was used in several approach-and-landing tests prior to the space missions.

OPPOSITE, BOTTOM: The other *Enterprise:* the model used in the production of the first *Star Trek* television series from 1966 to 1969, was donated to the museum by Paramount Pictures in 1974.

By the late 1980s, however, with the 50th anniversary of the end of World War II on the horizon, veterans and others were taking a very different view, arguing that the *Enola Gay* should be displayed with pride and asking why it remained disassembled and in less than ideal storage conditions. Reassembly and preparations for the presumed display of the bomber were already under way in 1987, when Martin Harwit took the helm.

In the rash of interviews marking his arrival, reporters indicated that the new director was "stirring things up." His basic intent was to shift the focus of the museum from its traditional role as a memorial toward a new emphasis on education, including the balanced consideration of the impact of science and technology on society. "What Martin Harwit is doing is taking his audience seriously," explained Roger Kennedy, the dynamic head of the National Museum of American History, adding that such a course was "a very different thing from presenting the history of technology as a succession of easy triumphs."

CAPTAIN'S MATERNITY UNIFORM Ca 1989 A United Airlines uniform marking a milestone in gender equity in the cockpit

The centerpiece of Martin Harwit's plan to shift the focus of the museum was to be an exhibition on strategic bombing. Such a project would, he explained to a *Washington Post* reporter, serve "as a counterpoint to the World War II gallery we have now, which portrays the heroism of the airmen, but neglects to mention in any real sense the misery of war." The new gallery would include a consideration of the nuclear dilemma. "We really are at a point now where, inadvertently or deliberately, it would be possible . . . to destroy the world," Harwit pointed out. Plans for such an exhibition, the reporter noted, had "so intrigued the Smithsonian community [that] it comes up in conversation with everyone from curator to museum director to the secretary of the Smithsonian himself."

As a means of setting the stage and educating themselves to undertake the exhibition, museum staff arranged an extraordinary public lecture series, in which authorities on every aspect of the subject, representing every point of view, discussed the history and impact of strategic bombing. In time, however, the availability of the restored *Enola Gay,* or as much of it as would fit in a museum gallery, led museum leaders to reduce the discussion of strategic bombing to a single unit in a smaller exhibition focusing on the atomic bombing of Japan.

The essential problem facing exhibit planners was clear. "As a single object," Samuel Batzli remarked, the *Enola Gay* "represents not only the heroes who created and flew the plane and hastened the end of World War II but also the victims of Hiroshima and the beginning of the atomic age." Considering the difficulty of achieving a "balanced" exhibition built around an object as emotionally laden as the *Enola Gay* within the context of the history of the museum, historian Alex Roland wondered "if the staff of the Air and Space Museum can

NCC-1701
Smithsonian

overcome this history and make the exhibit of the *Enola Gay* the educational opportunity it is planning."

A team of four members of the museum's Aeronautics Department completed work early in 1994 on the first draft of a five-unit script for an exhibition to be titled Crossroads, centered on the forward fuselage of the *Enola Gay* (it was not feasible to show the entire aircraft inside the existing museum building). Their aim was to tell the story of the atomic bombing of Japan and the end of the War in the Pacific. It would be an object-rich exhibition, with items ranging from crew artifacts to objects borrowed from Japanese museums illustrating the destruction on the ground.

Director Martin Harwit took the lead in enlisting a blue-ribbon advisory panel to guide the development of the exhibition, including a wounded veteran of the Pacific War, the historian of the Air Force, and several scientists and historians who were authorities in the field. While noting the need for some alterations, the group generally approved the initial script. At the same time, anxious to obtain broad comment on the project, the museum distributed copies of the draft script to a number of outside organizations.

The Air Force Association launched a sharp attack on the Crossroads script in April 1994. While noting that the museum team had covered the men of the 509th Composite Group that dropped the bomb "extensively and with respect," the organization left no doubt that it regarded the script as unpatriotic and "politically biased." In the association's view, the document gave very little attention to Japanese aggression and atrocities during World War II and too much attention to the effects of the atomic bombs and to later academic debates about whether they should have been dropped or not. Indeed, the draft script ultimately offended many parties across the United States, from veterans and their families to major media commentators and members of Congress, due to its narrow focus on the end of the war and the atomic bombings of Hiroshima and Nagasaki, at the expense of a full discussion of the Pacific War. Within months, the museum found itself at the center of a firestorm of criticism. The best efforts of Harwit and his curators to counter the tidal wave of opposition failed, as did extended attempts to negotiate a compromise script acceptable to all parties, using the American Legion as an intermediary. In January 1995 Smithsonian Secretary Michael Heyman canceled the exhibition after another outbreak of controversy over the number of casualties to be expected in an invasion of Japan. Finally, in May, Heyman asked for and received Martin Harwit's resignation. While the Secretary and other Smithsonian and museum leaders testified before a congressional committee, a new team of museum staffers produced a replacement exhibition that focused solely on the *Enola Gay* and her crew. It opened in June and lasted three years.

In the wake of the *Enola Gay* disaster, Heyman and the museum leadership conceded that the museum had made serious mistakes in the balance of the original exhibition and in the handling of key stakeholders. The Secretary asked the National Academy of Public Administration to review the museum's organization and management. The NAPA team finished their review in September 1995, finding fault with both organization and management and suggesting a less "academic" or "collegial management approach." Within a month, the Government Accounting Office issued a report critical of the museum's collections management policies, noting that the museum's collections management staff felt disenfranchised, citing management's emphasis on research and exhibits. Within the span of ten years, the museum that had once been criticized for conducting too little scholarly research was now criticized for spending too much time on scholarship.

OPPOSITE, TOP: Maj. James A. Ellison, base commander, reviews the flight line of the first class of Tuskegee cadets at the U.S. Army Air Corps basic and advanced flying school, Tuskegee, Alabama, 1941.

OPPOSITE, BOTTOM: An American MP guards an unfinished V-2 rocket at the underground plant at Nordhausen, Germany, shortly after the area was liberated in April 1945. Parts of several V-2s were used to assemble the rocket now on display at the museum.

BUTTERFLY HABITAT
1999
Duplicate of container flown on space shuttle mission STS-93 to evaluate metamorphosis in zero gravity

TURNING A CORNER

Robert Hoffman, then stepping down as Smithsonian undersecretary for science, became acting director while the search was under way for a new one. At the heart of the search was an explicit restatement of the museum's mission by an ad hoc committee of the Smithsonian Regents. The group insisted that the collections were of primary importance, and that exhibits, public programs, and research should be organized around objects in the collection. Moreover, a demonstrated capacity for leadership and strong management skills would head the list of qualities sought in a new director.

The search committee did not have to look too far. Retired Vice Adm. Donald D. Engen, the | to page 306 |

The Hubble Space Telescope orbits after the second servicing mission in 1997. In preparation for that mission, astronauts used the engineering test telescope in the museum to practice access to the spacecraft.

Because of its fragility, the backup mirror for the Hubble Space Telescope was the first artifact placed in the Explore the Universe gallery, opened since 2001. In the late 1970s Corning fabricated the 94-inch-diameter masterpiece of welded circular glass plates and strips, just a fraction of the weight of a normal solid disk of glass; Eastman Kodak optically ground, polished, and figured the mirror. Unlike the unit that went into space, it was perfect in every way.

HUBBLE SPACE TELESCOPE
BACKUP MIRROR

When the problem with Hubble's mirror was discovered after its launch in 1990, this mirror could have been used as a replacement if NASA had brought Hubble back to Earth for repair. But a far less costly, safer, and more creative solution was invented, and in 1993 corrective optics were installed into the existing mirror during the first servicing mission. The Hubble has been performing beautifully ever since.

And so the backup mirror remained in storage. When the museum expressed interest in displaying it, NASA officials had to determine if it still had scientific value and thus might have investigative use on other missions. Several observatories responded positively, but the cost of mounting the mirror so that it could work within Earth's gravitational field was prohibitive. So the museum acquired the mirror and a transport structure, and, courtesy of Kodak, it was transported to Washington, D.C., and set into place.

Standing before the mirror, one can see its full internal structure and support system. Many visitors ask the question "If it's a mirror, why can I see through it?" As a backup, it was not coated with a film of aluminum to make it reflective. A microscopically thin coating on the curved surface of the glass accounts for all the optical properties of the mirror, so it can collect as much light as possible and concentrate it into the smallest area possible: the essential capability of any great telescope.

In this stunning picture of the giant galactic nebula NGC 3603, the crisp resolution of NASA's Hubble Space Telescope captures various stages of the life cycle of stars in a single view. OPPOSITE: The backup mirror for the Hubble Space Telescope shows the honeycomb structure that gave it the resiliency it needed to be launched into space. The surface of the glass was never aluminized, allowing a view of the interior of the structure.

| from page 301 | museum's Ramsey Fellow in Naval Aviation History that academic year, was just finishing his book *Wings and Warriors*. In addition to his Navy experience, he had served on the National Transportation Safety Board and as head of the Federal Aviation Administration before coming to the museum. He was appointed in June 1996, and his first job was to reorganize the museum and execute a "reduction in force," an elimination of positions to meet the budget cuts that took place in the wake of the *Enola Gay* exhibition. Six months later, the Departments of Art and Culture and Astrophysics were gone, Education was severely cut back, and the film section of the Exhibits division was eliminated. The Aeronautics, Space History, Collections, and Archives divisions, and the Center for Earth and Planetary Studies remained intact, as someone had to tend to the core research and collections missions of the museum.

Even while the *Enola Gay* affair and its aftermath were unfolding, the museum had been the beneficiary of a great new opportunity to renovate Space Hall, the large windowed gallery that shows the backup Skylab space station, a test version of the Hubble Space Telescope, missiles and rockets in the Missile Pit, and the docked Apollo-Soyuz spacecraft. At the end of 1993, the industrialist and presidential candidate Ross Perot had purchased many rare Soviet space artifacts at auction, a by-product of the impoverishment of Russia's space program after the collapse of the Soviet Union in 1991. He almost immediately contacted the museum to find a place to show his artifacts, and he promised a generous donation from his foundation to make such an exhibition come about.

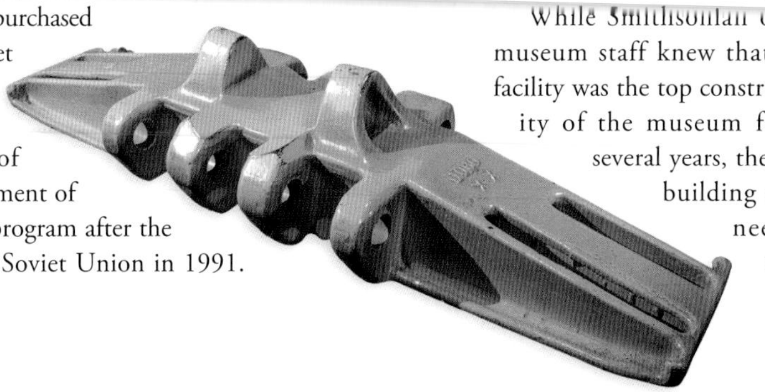

NASA CRAWLER TREAD
Mid-1960S
One of 456 treads used to transport the Saturn V rocket at one mile an hour from the assembly building to the launch pad

The result was the May 1997 opening of Space Race, the story of how the United States and the Soviet Union engaged in a missile and space competition that ultimately ended in cooperation in the post–Cold War era. Thanks to President Bill Clinton's declassification of the United States' first photoreconnaissance satellite, Corona, in 1995, the museum was also able to collect and exhibit one of the last surviving cameras from that series, providing an opportunity to discuss the secret intelligence dimensions of the Space Race as well.

Some of the Soviet artifacts, such as the backup spacesuits used by the first human space traveler, Yuri Gagarin, and the first person to walk in space, Alexei Leonov, are equally impressive, but smaller objects speak to the visitor too. Included in the exhibit is the arming key for Sputnik—the last piece of hardware taken out of the small spacecraft before its October 1957 launch and later fiery reentry. But perhaps the most unusual display is two slide rules, one used by rocket engineer Wernher von Braun for the United States, and one by Sergei Korolev for the U.S.S.R.—both made by the same German company.

While Smithsonian officials and museum staff knew that the Dulles facility was the top construction priority of the museum for the next several years, the 20-year-old building on the mall needed extensive repair. Surrounded by windows, the museum building is ideal for the public display of air and space artifacts, but not ideal for their preservation. But when seals leak, rainy days turn into indoor showers, and rust appears on hanging aircraft, it's time to fix the building. The entire wall of windows as well as skylights and doors needed to be replaced, and after an extensive discussion of whether to close the building and shorten construction time, or stay open with limited exhibits, the latter won out. For the next three years, from 1998 to 2001, the floor of the building was adorned with plywood, plastic covers on exhibits and artifacts, and temporary entrances that discouraged visitors. Unfortunately, Don Engen never got to see the end of the window-wall project. On July 13, 1999, he was killed in a glider crash in Nevada. Don Lopez became acting director, and another search began.

Relatively few aviation records were waiting to be shattered by the end of the 20th century. But on March 21, 1999, a bright red capsule with a 180-foot-tall balloon landed just north of Mut, in Egypt's Western Desert, establishing the record for the first nonstop round-the-world balloon flight. Bertrand Piccard and Brian Jones had been confined to the *Breitling Orbiter* for nearly three weeks, reaching more than 37,000 feet and at times having to chip ice away from the drinking water even inside the capsule. The gondola is now in a place of honor in the Milestones of Flight gallery. It would seem that the turn of the century was the time for long-distance balloon rides. In 2002, Steve Fossett succeeded in circumnavigating the globe solo in just under 15 days.

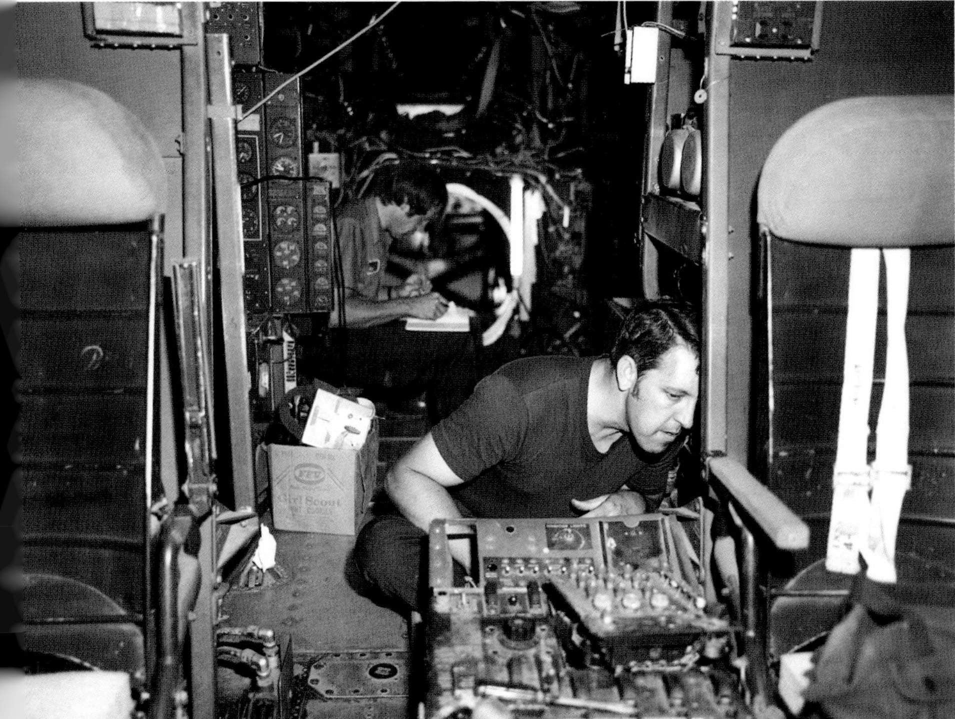

TOP: Birds' nests, grass, and fabric strewed the interior of the Boeing B-29 Superfortress *Enola Gay* after years of storage.

BOTTOM: Before the aircraft could be displayed in the museum, restoration of the cockpit and interior of the *Enola Gay* was intensive.

His *Bud Light Spirit of Freedom* gondola is in the Pioneers of Flight gallery, adjacent to many other firsts.

A NEW MILLENNIUM

January 1, 2000, was Gen. John R. "Jack" Dailey's first day as director of the National Air and Space Museum. He had been training for this job throughout his career, as Marine aviator, a commander of several aircraft wings and programs, assistant commandant of the Marine Corps, and deputy associate administrator of NASA following his military retirement. His first job at the museum was to complete the funding and construction of the Dulles extension, assisted by Don Lopez, who had returned from retirement to become Don Engen's deputy director. He retained that position. The story of Dulles is the subject of the next chapter, but while that extensive effort was ongoing, the museum on the mall continued on many fronts, requiring further funding.

Tapping into corporate philanthropy is no easy task. Only a very few individuals are capable of and truly energized at the prospect of making a significant donation to the Smithsonian, and corporations look for something different. Following extensive discussions, the museum closed down the Langley IMAX Theater one night in 2002, reopening it the next morning as the Lockheed Martin IMAX Theater. As had now become customary for the museum, such a deed did not go unnoticed. One member of Congress wrote, saying how dare the museum remove Secretary Langley's name from the theater and substitute instead a commercial entity. | *to page 310* |

Bobbe Dyke, a former mathematics teacher who traveled extensively with her husband during his service in the Air Force, moved to the Washington, D.C., area in 1964. In early 1976, she became a museum docent, and she was at the museum on the day when the mall building was officially opened. She works at the museum to this day.

BARBARA W. "BOBBE" DYKE

ON BEING A DOCENT

Some time in 1975, my husband and I were having dinner with two old friends, Don and Glyn Lopez, and I mentioned that I was going to apply to be a docent at the Hirshhorn, which had just opened.

"What do you mean you're going to be a docent at the Hirshhorn?" Don said. "Air and Space is going to open next year." The next day I got a telephone call from someone at the museum, saying, "I understand you want to be a docent at Air and Space. When can you come in?"

The docent experience is ever changing and often rewarding. I have given more than 3,400 tours in my more than 34 years at the museum, and each one has been different. I estimate I have talked to about 70,000 people during that period. That seems like an awesome responsibility.

I have toured three kings, a sultan, two princes, three first ladies, one president, generals and admirals, ambassadors, congressmen and senators, astronauts and cosmonauts, a patriarch, and CEOs. I've also toured children, probably more than 15,000 in all. There have been rich kids and kids from the ghetto. Some were interested; some were not. But I think I was able to inspire some of them. Perhaps I made a difference in a few of their lives.

One busy summer day at NASM, a number of years ago, I happily agreed to give a tour to a person identified to me in such a way that I thought it would be Robert Redford. It was sometime later that I learned the tour was not for the blue-eyed actor, Robert Redford, but rather for a red-eyed machine named Robot Redford.

Robot Redford, who strongly resembled R2D2 of Star Wars *fame, was a remotely controlled automaton that had been created in order to demonstrate to NASA some of the latest advances in robotics. I was surprised by the depth of his knowledge until I realized that his operators, easily lost in the crowd, were usually standing where they could read the appropriate labels and then quiz me accordingly. Needless to say, the large group of tourists we attracted enjoyed Robot's many quips and were amazed at the very idea of a robot being a fellow tourist.*

When I have spare time in the museum, I sometimes stand on Skylab's back porch. I enter into conversation with visitors and give out information until I back up the line too much. Recently, a woman I encountered said, "This is the sixth time I've gone through Skylab, and it's the first time I've understood anything about what I was seeing—because you were there."

That's what docents do. We help make the exhibits come alive for visitors.

| from page 307 | He requested a reversal. The museum did not comply, but it did add Samuel P. Langley to the name of the Early Flight gallery in the museum. Of course, when you look beyond the corporate name, the names of Allen Loughead (the original spelling), or Glenn L. Martin, or William E. Boeing, and many others are enshrined not just in corporate names, but in the history of aerospace. Perhaps what that congressman didn't realize was that the funds from that naming would help preserve the objects in the plywood box that was opened four years later in preparation for the Udvar-Hazy Center. In it was Secretary Langley's personal lab equipment for measuring lift and aerodynamic forces, part of his quest to build the first heavier-than-air manned aircraft—covered with nearly a hundred years of dust.

The first major exhibit opening under Dailey's tenure was Explore the Universe in 2001. The old Stars exhibit gallery, dating from 1983, was never satisfying to the museum, to visitors, or to the curators. It was crowded, and it displayed a great deal of "star" culture (the Texaco star) and not enough science. To correct that, and in response to the swarm of astronomical satellites that had been launched, a new project to build an astronomy gallery had begun in the early nineties by astrophysicist Harwit. After a long gestation, Explore the Universe was funded by TRW and the National Science Foundation. It includes some rare historical objects, like the telescope tube of 18th-century British astronomer William Herschel and the Newtonian cage from the hundred-inch telescope on California's Mount Wilson. The gallery describes how the changing tools for human observing—from naked eye to the telescope, the camera, the spectrograph, and the digital instrument—have changed our understanding of the universe. The exhibit includes the backup mirror for the Hubble Space Telescope (the one without the flaw, as opposed to the one that is in space).

OPPOSITE, TOP: Museum scientist Andrew Johnston surveys dunes near Death Valley in 2005 to understand similar dune structures on Mars.

OPPOSITE, BOTTOM: Robert Craddock from the museum's Center for Earth and Planetary Studies points out the surface of Mars as seen through 3-D stereo glasses on the museum's Mars Day, September 16, 2000.

CORONA FILM RETURN CAPSULE 1972
Recovered in midair from the last Corona photoreconnaissance satellite mission

With an opening planned for September, the walk-through for the director and senior staff was at 9 a.m. on September 11, 2001. Within the next few hours, the staff was either watching smoke rise from the airplane crash into the Pentagon or joining thousands walking or attempting to drive out of the city. Those who stayed behind listened to conflicting accounts on the radio all morning, and wondered whether four blocks from the U.S. Capitol was a safe place to be. Given the events of the fall, the opening of Explore the Universe understandably made hardly a ripple in the media. Even so, the gallery became very popular, attracting an average of 3,000 people per day nearly a decade later.

The year 2003 was the centennial of the Wright brothers' flight, and not surprisingly the museum was centrally involved. Director Jack Dailey was chair of the Centennial of Flight Commission, and President Clinton appointed Tom Crouch chair of the Centennial Advisory Committee. A team of designers, educators, and curators, led by Peter Jakab, was putting the finishing touches on a grand new exhibition, The Wright Brothers & the Invention of the Aerial Age, which featured the 1903 Wright Flyer, the world's first airplane, as the centerpiece. Displayed at eye level for the first time since the preservation effort of the mid-'80s, the Flyer was surrounded by an exhibition that included one of only five Wright bicycles, pieces of wood and fabric from the original Flyer that went to the moon on Apollo 11, works of art, and a host of other items that helped visitors understand the Wright achievement and its impact on the world. Originally planned for a two-year display, the exhibit proved so popular that it remains in place to this day.

In 2007, Aeronautics curators finally got their long-awaited wish—the complete renovation of the Hall of Air Transportation—this time with a theme and a story to go with the artifacts. America by Air distilled curatorial research on the growth of commercial aviation to tell the engaging story of the U.S. airline industry. An excellent design incorporating leading-edge educational interactives was combined with the same beautiful hanging aircraft that had been in the museum for three decades—the Boeing 247-D, the DC-3, and others. But added was a much bigger treatment of the jet era. After extensive negotiations with Northwest Airlines, one of the first Boeing 747s used for commercial transport was donated to the museum. The cockpit and forward fuselage was cut apart in North Carolina; shipped to Florida, where it was reassembled and repainted; then moved to Washington in nine pieces for assembly on the wall of the gallery. Today, visitors can walk through not only a Skylab space station in Space Race, but also the nose of a mammoth 747 in America by Air.

Meanwhile, research has flourished, hardly affected by changes of leadership

and claims of too much or too little support for scholarship. Nearly 200 books have been published by museum staff, including prize-winning biographies of the Wright brothers and Wernher von Braun, an award-winning history of the Hubble Space Telescope, several groundbreaking works on the history of aerodynamics, and literally hundreds of professional papers in the historical and scientific journals. The Center for Earth and Planetary Studies, starting from the analysis of Apollo mission data in the 1970s, now has team members on all active missions to Mars (the Mars Exploration Rovers, Mars Reconnaissance Orbiter, Mars Express spacecraft), to Mercury (MESSENGER spacecraft), and the moon (Lunar Reconnaissance Orbiter). During shifts as lead scientist for the Mars rovers Spirit and Opportunity, geologist John Grant helps select paths for them from his third-floor office.

Now, on any given day, someone is refinishing the spar of an 80-year-old airplane, someone is analyzing the composition of a 40-year-old spacecraft, someone has her head deep in the archives looking at the history of an aerospace pioneer, someone is looking at the latest results from a Mars probe, someone is on the phone with a prospective donor, someone has just changed the design of the next exhibit gallery, someone is trying to figure out how to make the budget last until the end of the year, and someone is refiling the archival records that were brought out for a visiting researcher. Meanwhile, in the downtown museum alone, an average of 20,000 people see only the tip of the iceberg, helped along by volunteers and educators who distill our air and space heritage for millions each year.

In 2002, researchers at the museum's Center for Earth and Planetary Studies discovered evidence for an ancient lake on Mars that was the source of one of the major river valleys. The lake would have been more than twice the area of the Caspian Sea, the largest lake on Earth.

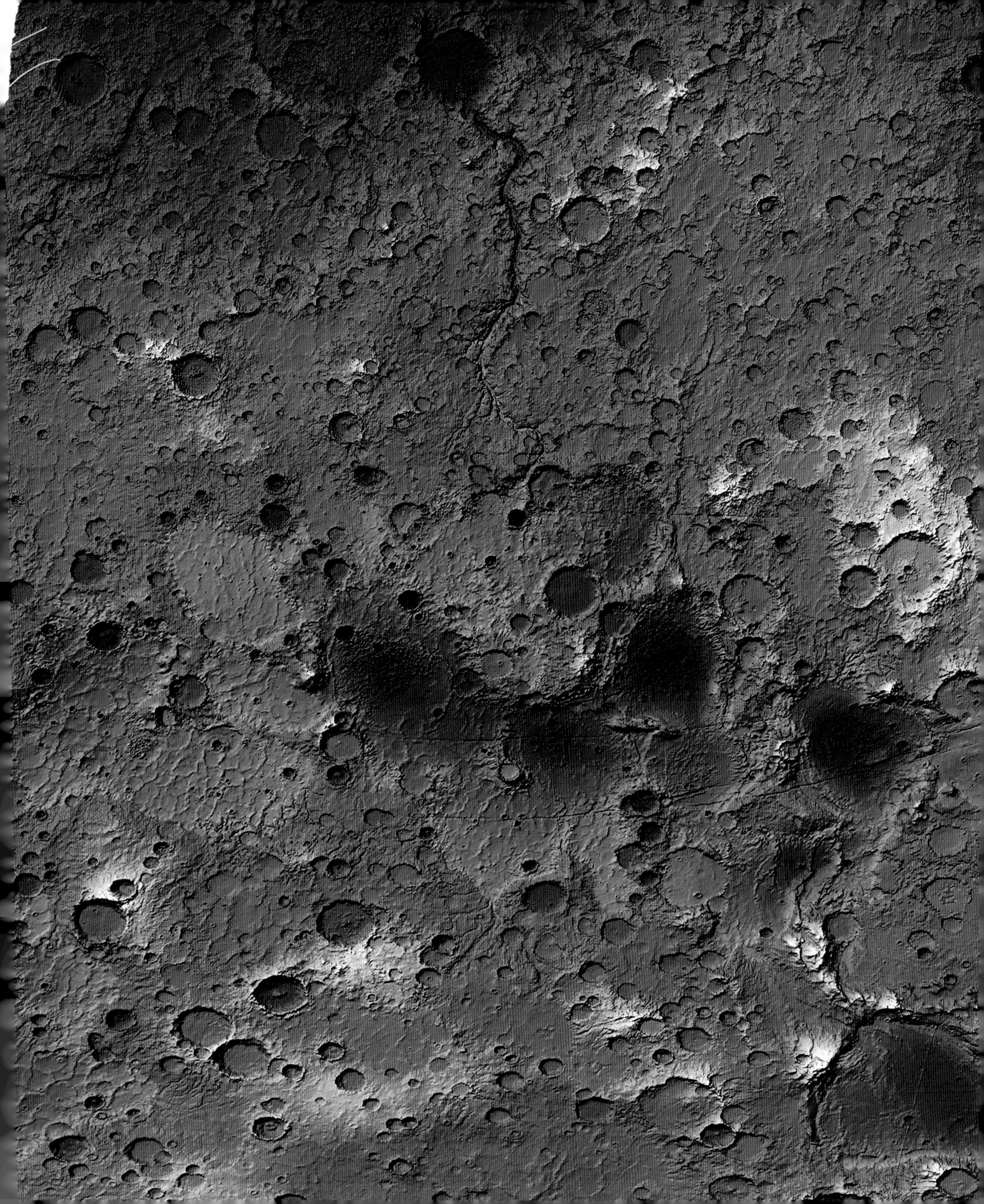

COLLECTIONS

Approximately 1,200 pressure suits, gloves, helmets, boots, and attachments make up of the most significant collections of the National Air and Space Museum. It is the most comprehensive spacesuit collection in the United States.

SPACE SUITS

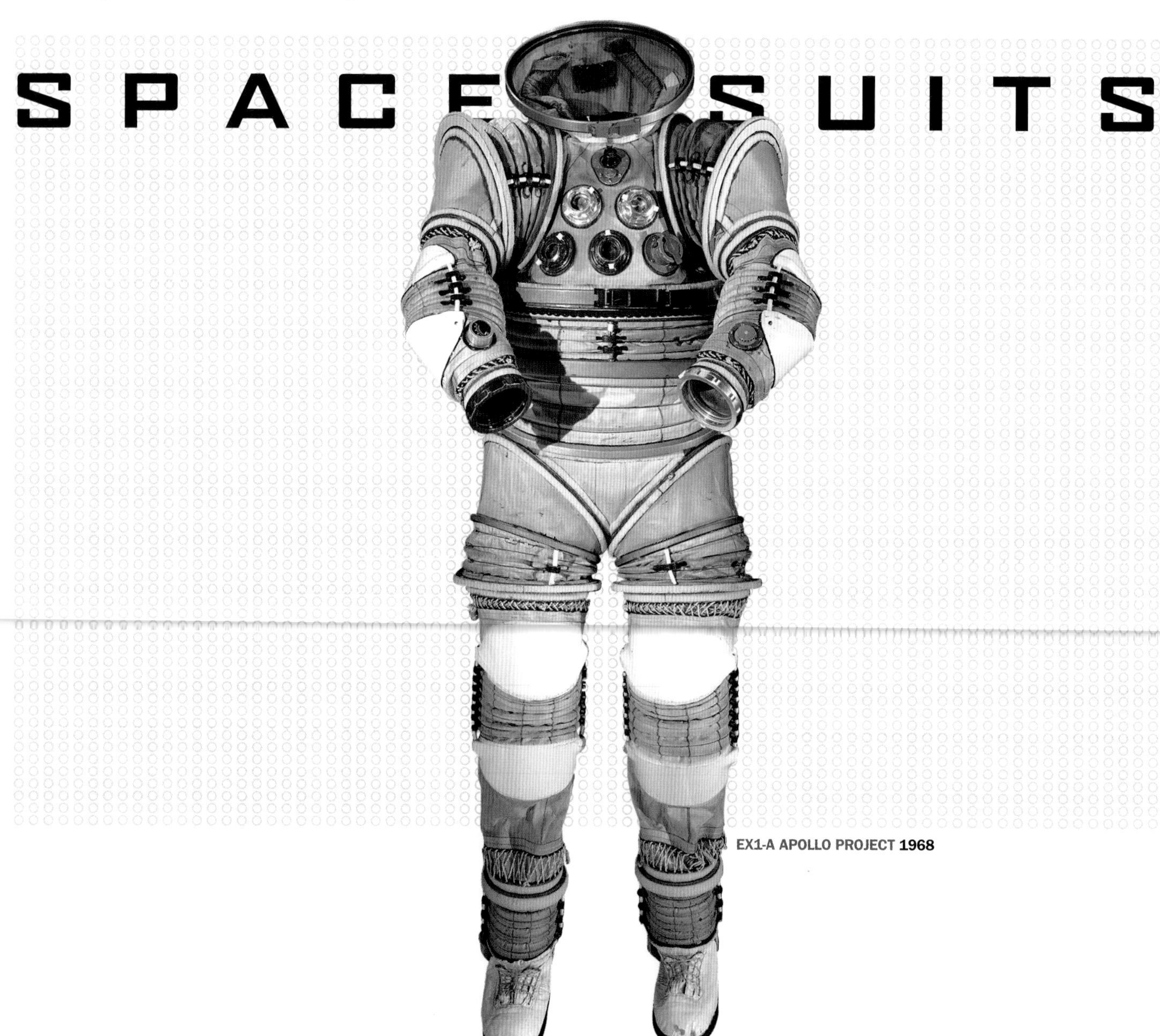

EX1-A APOLLO PROJECT **1968**

G2-G - SHEPARD **1962**

SOKOL KV-2 **1995**

RX-3 PROTOTYPE **1965**

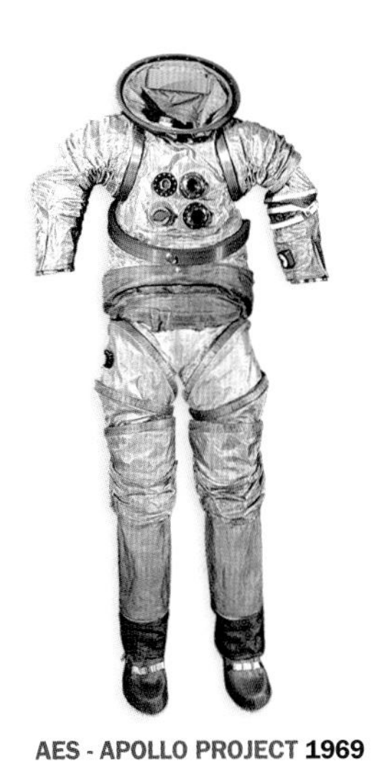

AES - APOLLO PROJECT **1969**

MARK V MODIFIED **1968**

GAGARIN VOSTOK I TRAINING **ca 1960**

GLENN MERCURY SUIT **1961**

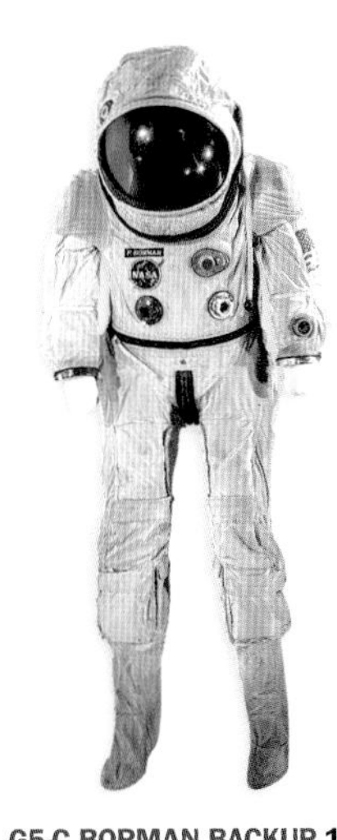

G5-C BORMAN BACKUP **1965**

A7-LB-ASTP - STAFFORD **1974**

A5-L APOLLO PROTOTYPE **1965**

MARK IV **ca 1965**

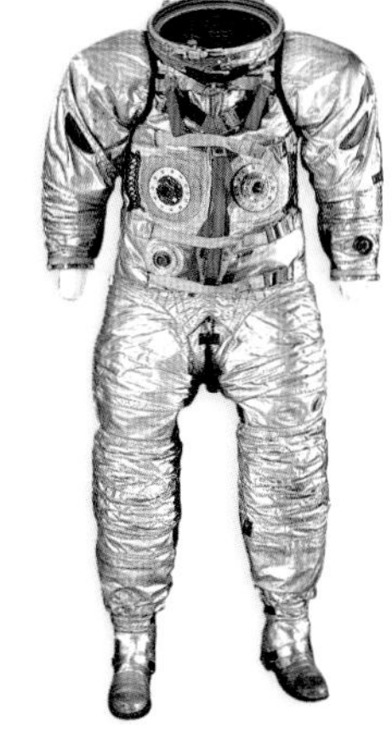

A4-H APOLLO DEVELOPMENTAL **1964**

MARK V MODIFIED **1968**

SPD-143 DEVELOPMENTAL AX2-1 **1963**

XN-20 APOLLO PROTOTYPE **1964**

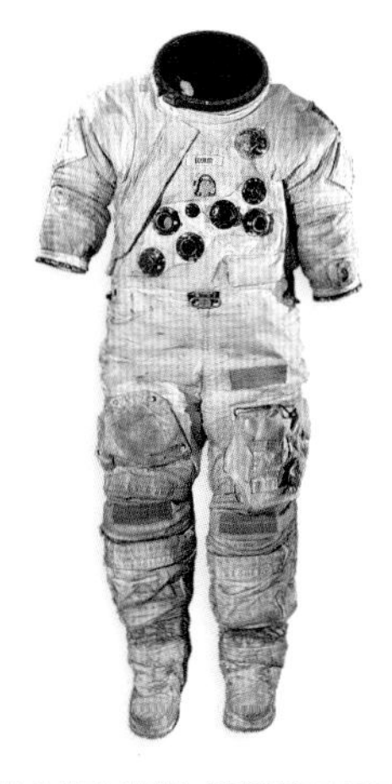

A7-LB EVA SUIT - SCHMITT **1972**

A7-LB EVA SUIT - YOUNG **1971**

A7-LB EVA SUIT - SHEPARD **1970**

A7-LB EVA - BEAN **1973**

S-1022 APOLLO ADVANCE CONCEPT **1967**

MH-5 FIRST MOL TRAINING **1967**

RX-4 **1966**

G2-C - SLAYTON **1964**

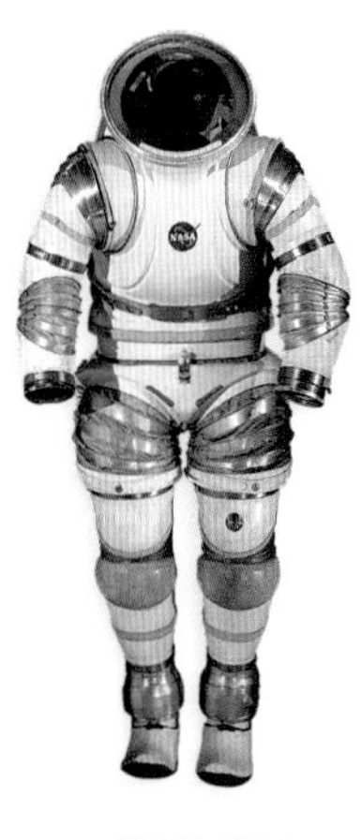

RX-2A **1964**

MARK II - MODEL "R" **1956**

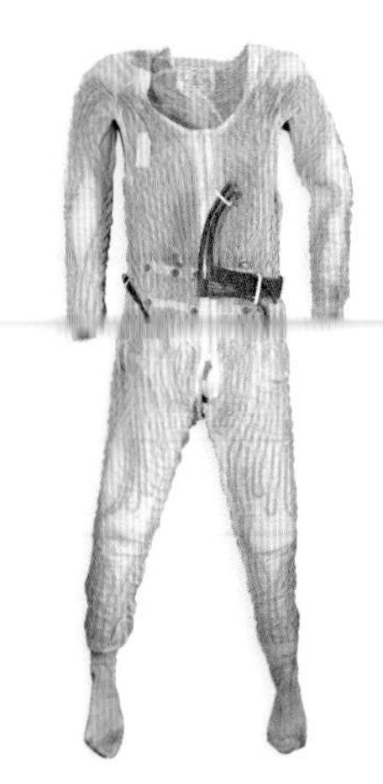

LIQUID COOLING GARMENT **1968**

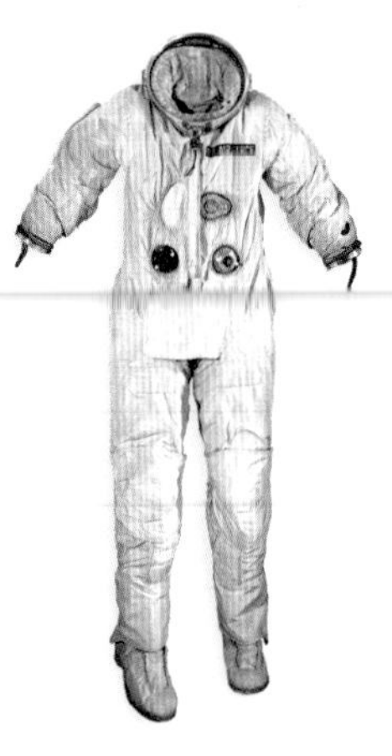

MD-1 MOL PROTOTYPE **1966**

AX-3 **1974**

MARK II - MODEL "O" **1956**

RX-2 APOLLO APPLICATIONS **1963**

The first spacesuits were modified high-altitude aviation pressure suits. Sealed life-support systems turned them into individual spacecraft. As astronauts performed increasingly sophisticated tasks, the suits had to balance the demands for mobility and protection in open space. These suits have layers of materials to protect the astronauts from temperature extremes, micrometeoroids, and radiation while maintaining dexterity for ease of motion. Over time, fabrics and metals reacted to one another or deteriorated. In 2001, the museum received a Save America's Treasures grant to understand the decay and preserve these artifacts for future generations.

S-1032 LAUNCH-ENTRY SUIT **1989**

4.5 THE SMITHSONIAN AND THE *ENOLA GAY*

After Col. Paul Tibbets (USAAF) and his crew had successfully delivered the first ever atomic bomb used in war, the Boeing B-29 Superfortress *Enola Gay* became famous. The bomber and crew had helped end World War II and had initiated the nuclear age.

The Boeing B-29 Superfortress *Enola Gay* is an icon of the National Air and Space Museum collection, an object that evokes immediate recognition and an emotional response from visitors. However people react to the airplane that dropped the atomic bomb on Hiroshima, visitors know that they are in the presence of a machine that changed the course of history. It is also an object that has been closely intertwined with the history of the museum ever since its founding in 1946.

Born as a 1939 Boeing proposal for a very long range bomber, the B-29 was the ultimate in piston-engine aircraft technology. By the time production ended in 1946, 3,970 Superfortresses would roll off the assembly lines. From design to deployment, the B-29 project was personally supervised by Gen. Henry H. "Hap" Arnold, the commander of the U.S. Army Air Forces. Arnold was certain the Superfortress would prove airpower advocates' contention that the long-range strategic bomber was the ultimate war-winning weapon.

The B-29 program stretched contemporary aeronautical technology almost to the breaking point. The Air Forces demanded a 5,000-mile range, which necessitated pressurized crew compartments for extremely high altitudes—the higher the altitude, generally, the greater the range. As a result, most of the defensive gun turrets had to be remotely aimed and fired by the crew. Four Wright R-3350-23 turbo-supercharged radial engines, each developing 2,200 hp, propelled the bomber. With its twin rows of 18 cylinders, and magnesium components that could burn uncontrollably once ignited, the engines were prone to overheating and catastrophic fires. More B-29 crews were lost to accidents and malfunctions than were lost in combat.

The first B-29 raid on Japan was launched from Chengdu, China, on June 15, 1944. But basing the aircraft in China proved difficult, and so the bulk of the B-29 buildup was shifted to the Mariana Islands in the Pacific. Nor did the traditional doctrine of

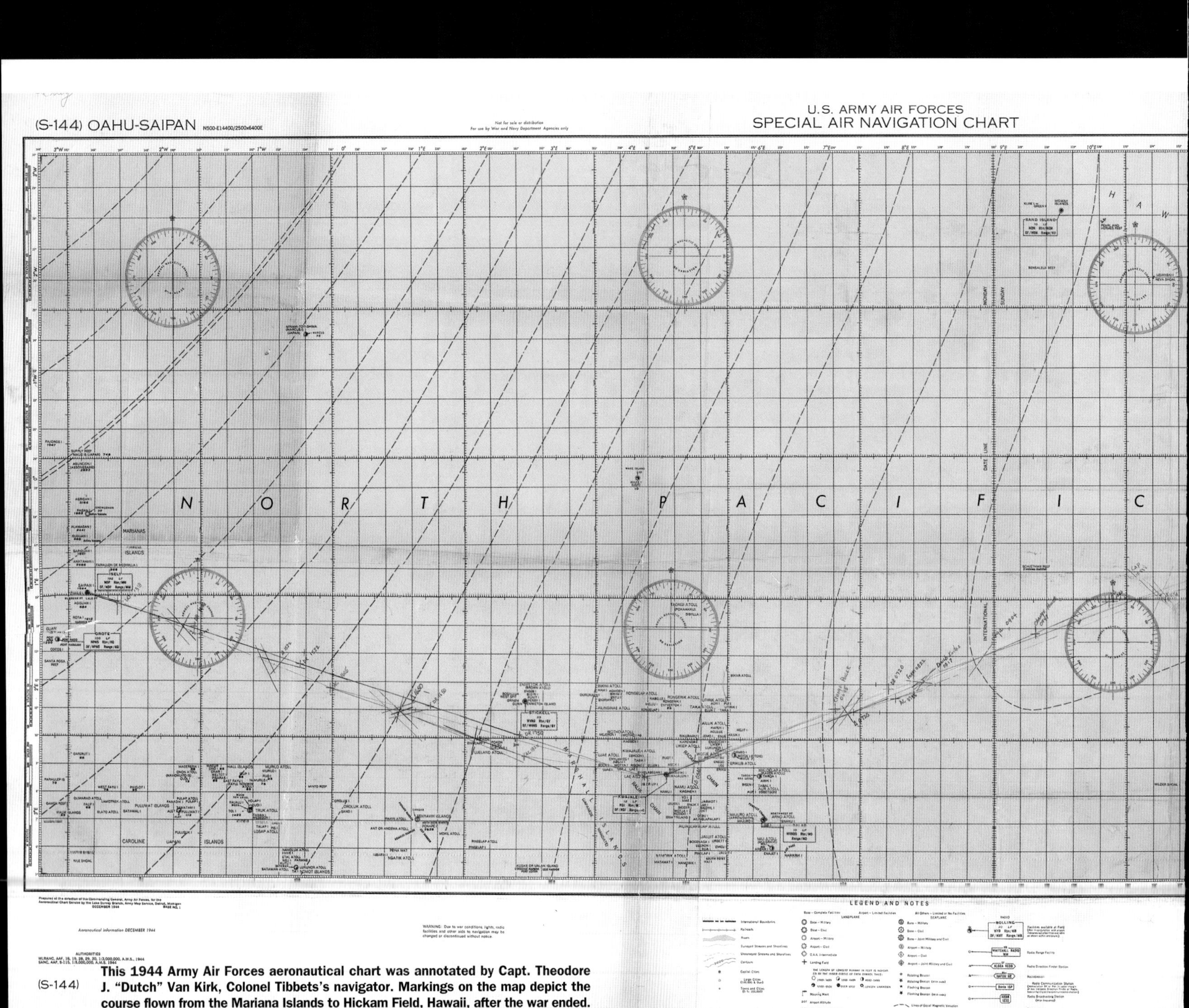

This 1944 Army Air Forces aeronautical chart was annotated by Capt. Theodore J. "Dutch" Van Kirk, Colonel Tibbets's navigator. Markings on the map depict the course flown from the Mariana Islands to Hickam Field, Hawaii, after the war ended.

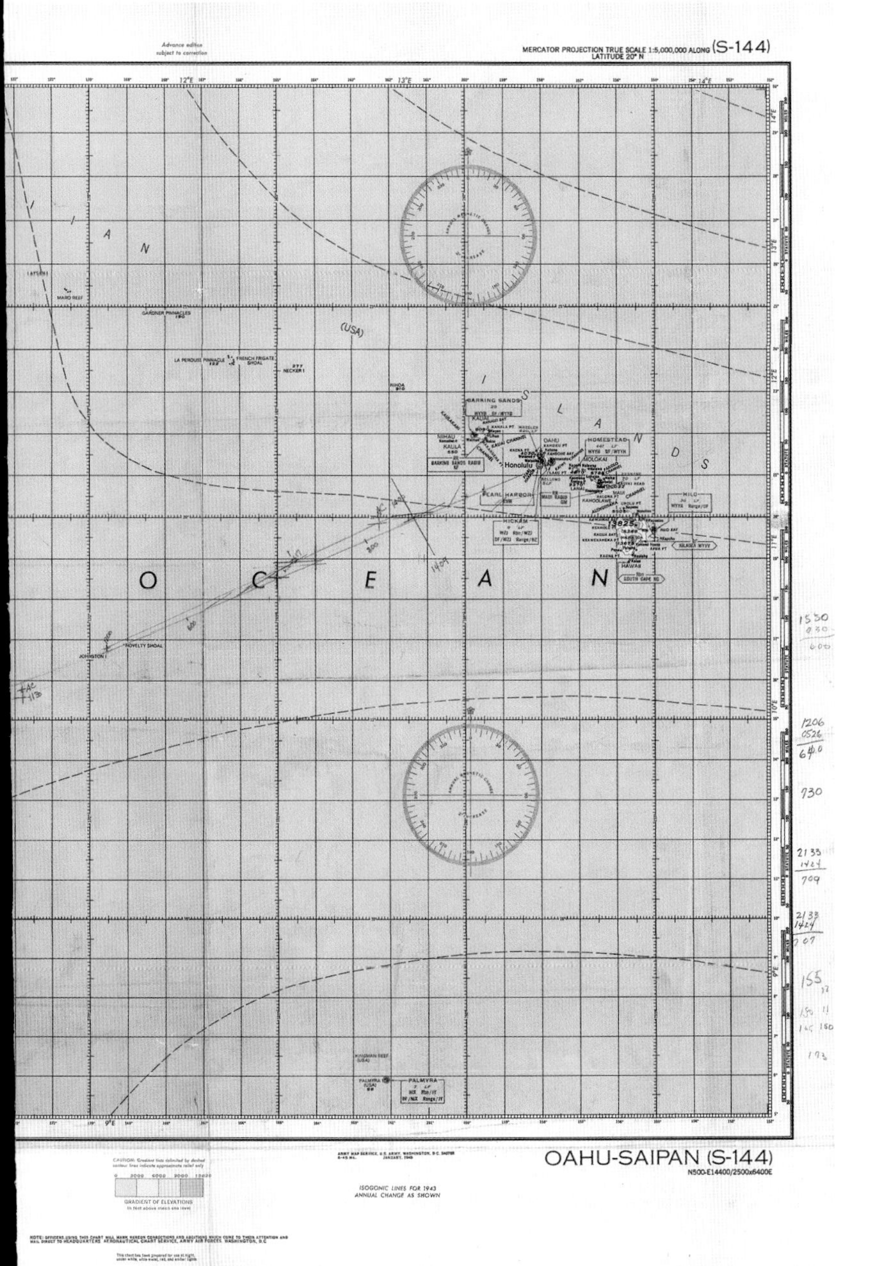

high-altitude precision bombing work well, in part because of jet-stream winds over Japan.

When Gen. Curtis LeMay took over the Marianas command in January 1945, he began to consider a radically new approach. For a raid on Tokyo on March 9–10, he stripped most of the defensive armament from his B-29s and sent them in at night and at low altitudes (5,000–9,000 feet), primarily armed with incendiary bombs. The resulting firestorm killed tens of thousands of civilians and incinerated 16 square miles of the city. In the weeks that followed, the combination of low-altitude incendiary attacks and the use of B-29s to heavily mine Japanese home waters drove city dwellers into the countryside, brought the nation close to starvation, and drastically reduced the enemy's productive capacity.

As early as fall 1943, the B-29 was identified as the only American aircraft capable of conducting atomic attacks. Between February 1944 and December 1946 a total of 65 B-29s were specially modified under the code name Silverplate. The USAAF activated the 509th Composite Group, the unit that would carry out the atomic attacks, at Wendover, Utah, on December 17, 1944, with Col. Paul W. Tibbets commanding. The first of its B-29s arrived at North Field, Tinian Island, on May 18, 1945. On August 6, Colonel Tibbets flew the B-29 *Enola Gay,* named in honor of his mother, on the first atomic mission, carrying the uranium gun bomb nicknamed "Little Boy." It devastated the city of Hiroshima.

Three days later, on August 9, Maj. Charles W. Sweeney's crew flew another Silverplate B-29, *Bockscar,* on a mission that hit Nagasaki with the plutonium implosion bomb "Fat Man." Although a more powerful weapon, the bomb detonated nearly two miles from the intended target and the blast was mitigated by hilly terrain. The previous day the Soviet Union had declared war on Japan. In the wake of this triple blow, Emperor Hirohito announced the surrender of Japan on August 15. The war ended without the need for a costly invasion of the Home Islands.

Enola Gay returned to the United States from Tinian in November 1945 and was temporarily stationed at Roswell, New

Shortly after Colonel Tibbets delivered the "Little Boy" atomic bomb over Hiroshima on August 6, 1945, a reconnaissance plane took this image from 25,000 feet above the city. RIGHT: Restored to original condition in the late 1990s after hundreds of man-hours, the front cockpit of the B-29 *Enola Gay* was manned by pilot Col. Paul Tibbets (left seat), bombardier Maj. Tom Ferebee (center nose seat), and co-pilot Capt. Robert Lewis (right seat) on August 6.

Mexico. In April 1946, Tibbets flew the aircraft back to the Pacific as part of Operation Crossroads, the atomic tests at Bikini atoll. In July, Tibbets took the plane to the United States again, landing in Tucson, Arizona. The *Enola Gay* was dropped from the Air Forces' inventory the following month.

At the end of World War II, General Arnold had ordered the collection and storage of U.S. combat types, along with as many captured enemy aircraft as could be found. He specifically listed the *Enola Gay* for inclusion in the aircraft bound for a National Air Museum, a facility that would be a "permanent shrine" devoted to American aeronautical achievements.

In 1949, Colonel Tibbets flew the *Enola Gay* from Tucson to Park Ridge, Illinois, where the Smithsonian officially took ownership. When that facility was closed in 1952 as a result of Korean War requirements, the B-29 was flown to Pyote, Texas, for storage. The last flight of the *Enola Gay* came on December 2, 1953, when the aircraft landed at Andrews Air Force Base outside Washington, D.C., where it sat outdoors for the next seven years. Unfortunately, damage from weather, vandalism, theft, and animal intrusions were the inevitable result, but the Smithsonian lacked the resources to do anything else. Only after Paul Garber had built up the Silver Hill storage facility nearby was it possible to disassemble the historic craft in 1960–61 and truck it there, where *Enola Gay* remained until restoration began in 1984.

During that quarter century, the airplane was the subject for both symbolic use by the museum and criticism by outsiders. It appears in the architectural renderings and in public statements for a new NAM building in 1955, but was not singled out so clearly when the next attempt was made in 1964–65. Perhaps the museum leadership was simply aware that *Enola Gay* now required an expensive and time-consuming restoration, or perhaps it was already becoming uncomfortable with the controversial character of the Hiroshima bombing, at least in some quarters.

The rising tide of antimilitary feeling that peaked during the Vietnam War certainly had an impact. Some museum officials

CLOSED
CLOSED
CLOSED
CENTERING
TURN COMPENSATION
TURBO BOOST SELECTOR
BOMB DOORS
WARNING

expressed opposition to showing the artifact in newspaper stories from the mid-'60s to the early '70s, and even Republican Senator Barry Goldwater, an Air Force Reserve general, said that it should not be exhibited in the new building he championed in the early 1970s. In any case, the downsizing of the museum to fit the budget effectively meant that there was no public space big enough for the fully assembled aircraft when the opening finally came in 1976.

Along with praise for the outstandingly successful new monument to aerospace technology came academic criticism that the museum was deliberately running away from difficult topics like strategic bombing and the atomic attacks on Japan. Why not show the *Enola Gay?* was a key question of the liberal critics, and they were joined in that by the veterans of the 509th Composite Group and other units—for entirely different reasons. As the museum began formulating plans for a Dulles center in the 1980s, the B-29 again served a useful symbolic purpose, as one of the key aircraft that could be shown there. Museum Director Walt Boyne initiated the restoration of the *Enola Gay* in 1984, and Garber Facility craftspeople pulled the forward fuselage into the shop.

The effort to restore the aircraft shifted into high gear with the approach of the 50th anniversary of the end of World War II in 1995. The new director, Martin Harwit, hoped to use it in a major exhibition on strategic bombing. After the impracticality of exhibiting the *Enola Gay* inside or next to the mall building was confirmed, Harwit decided in late 1992 to display the first 60 feet of the fuselage, which had the evocative name painted on the nose, in an exhibition on the atomic bomb and the end of the war. That led to the well-known public uproar in 1994–95, the cancellation of the exhibit, and Harwit's resignation (see Chapter 4). The aircraft's forward section and lesser components became the centerpiece of a smaller, less controversial display that opened in June 1995 and lasted three years.

Meanwhile, restoration continued on other parts of the aircraft, now with the objective of showing it at Dulles. It was so massive a job that Garber's able craftspeople tallied 300,000 work hours during the nearly two decades of restoration. Finally, in spring 2003, they trucked the *Enola Gay* in pieces around the Washington Beltway to the new Steven F. Udvar-Hazy Center.

B-29s were not built to be disassembled once they left the factory. Putting one back together without a two-story assembly building proved a monumental challenge for the museum's restoration professionals. The first step was to assemble the fuselage and wing roots. Next, the crew lowered the wing box into place and installed the eight main bolts that connect the two structures. Detailed work, including the installation of the four engines, continued for the next six months. Finally the nearly completed plane was delicately rolled onto the three large platforms that then lifted the craft to its display height nearly eight feet above the floor, providing more floor space for World War II aircraft and allowing visitors the best view of the cockpit interior from the second-level catwalk. The complete airplane went on display for the first time when the center opened to the public on December 15, 2003, accompanied by a small anti-nuclear protest (see Chapter 5).

The *Enola Gay* offers rich opportunities for storytelling. As a B-29, it provides the ideal means of presenting the story of wartime aircraft production, aircrew training, weapons production, the relationship between the military and the aviation industry, and the impact upon the American economy and society during and after the war. The *Enola Gay* also serves as a direct link to the Manhattan Project—which produced the atomic weapons used against Japan. These two projects were the most secret and the most expensive of any U.S. military projects during World War II. The many personalities involved make interesting history and, potentially, exciting exhibit material.

For the time being, these stories await expanded exhibit capability at the Udvar-Hazy Center. But even without them, the fully assembled aircraft remains one of the true highlights of the national collection. *—Dik A. Daso and Tom D. Crouch*

At the Udvar-Hazy Center after reassembly in the fall of 2003, the B-29 *Enola Gay* awaits installation on stands that lift the plane nearly eight feet above the concrete floor. Additional lighting makes it possible for visitors to view the cockpit interior from the second-level catwalk near the nose of the bomber.

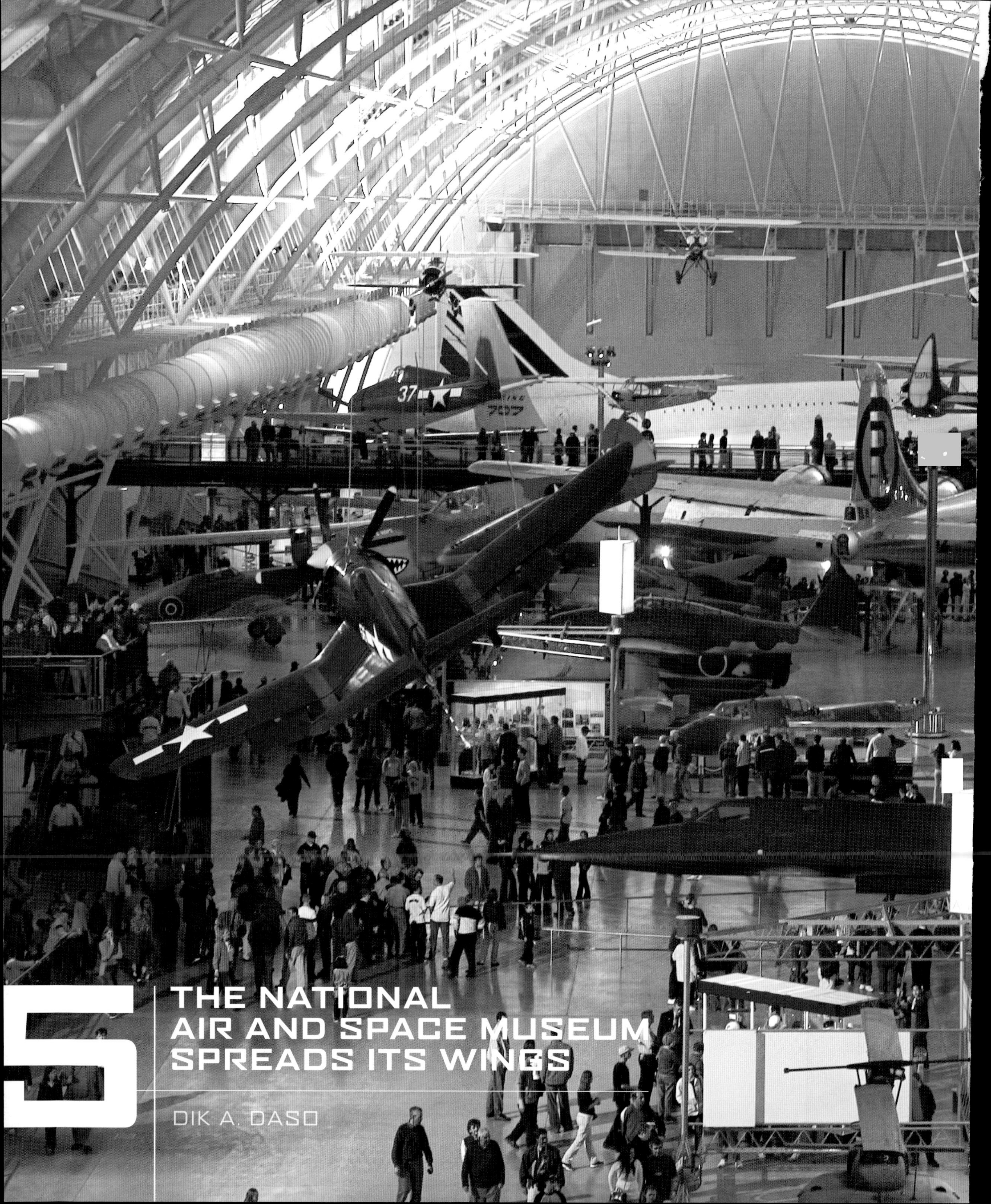

5 THE NATIONAL AIR AND SPACE MUSEUM SPREADS ITS WINGS

DIK A. DASO

44
44

Several years before the museum on

it became obvious to staff that the space would not be sufficient to hold the entirety of the air and space artifacts that had already been collected at the Silver Hill facility in Maryland. The original pre–Vietnam War design drawings were for a building that was larger and more spacious than the one that budget constraints of the early 1970s allowed. But a smaller building on the mall was far better than none. Even as the very first shovel of D.C. clay was being removed from the current museum site, it was a foregone conclusion that another facility would one day be needed to store and display not only the larger artifacts—like the Boeing B-29 *Enola Gay*—but also smaller craft needing complete restoration, which might not be ready for decades *the National Mall opened its doors*

to come. / In the 1960s groups outside of the Smithsonian system had suggested that a location permitting expansion and the construction of an enormous display facility would be beneficial to the rapidly expanding air and space collection. Some tourism interests from the Commonwealth of Virginia, for example, envisioned a tremendous growth in visitors from a Dulles Airport–based air museum complex. An active campaign was directed at Dulles decision-makers to pursue such a project. Bureaucrats deflected such inquiries by claiming that the Dulles Airport Master Plan and expansion designs would not permit such a facility's construction nearby—too much highway traffic, too much interference with airport operations.

Bell XV-15 Tilt Rotor
1979

Loudenslager Laser 200
1975

FROM THE MUSEUM'S COLLECTION

1960s TO 2000s

Virgin Atlantic Global Flyer
2005

Cosmos Phase II
1992

Rockwell Shrike Commander
1968-1979

Sukoi SU-26M
1983

Grumman F-14 (R) Tomcat
1974

Republic F-105D Thunderchief
Ca 1965

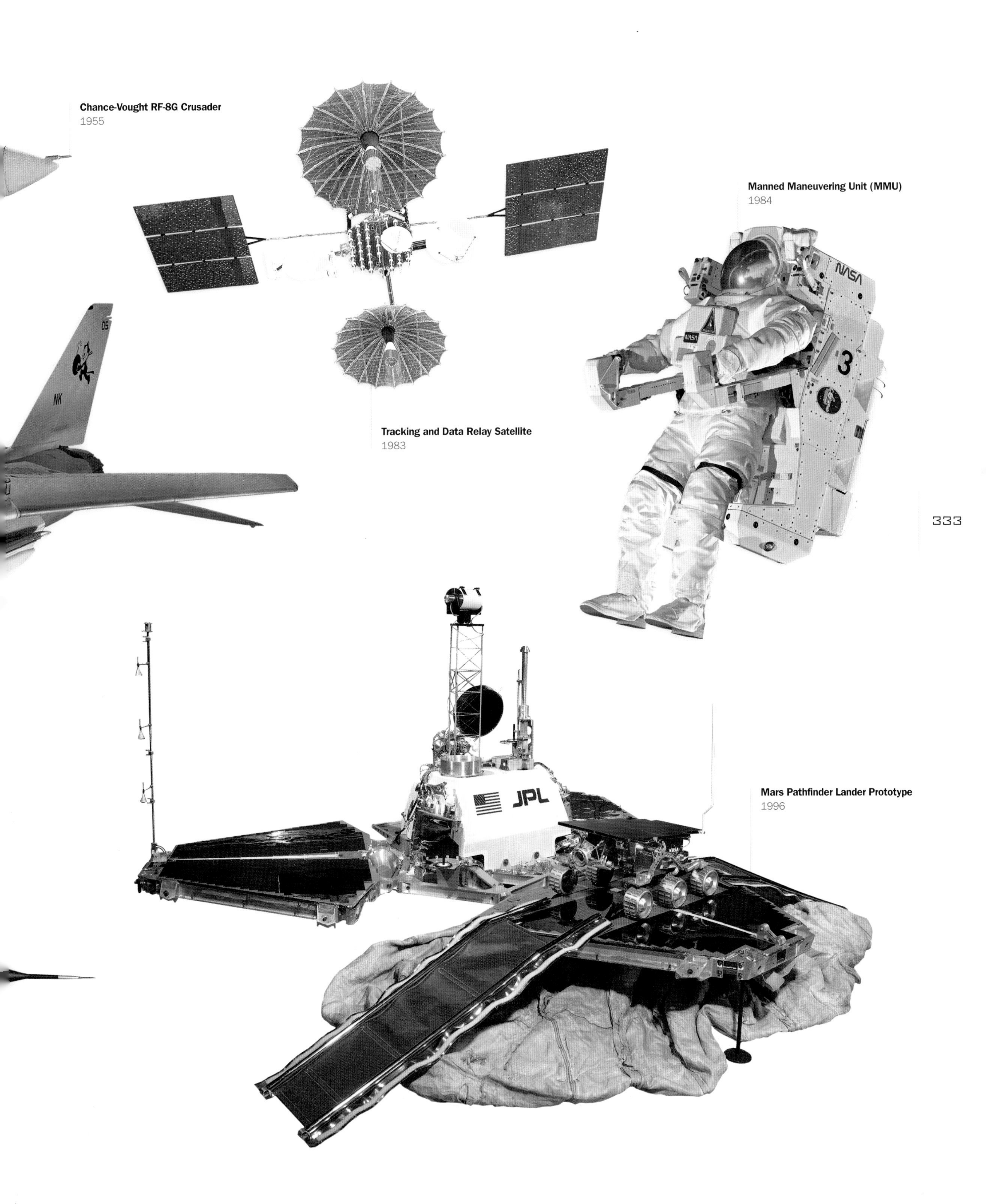

Chance-Vought RF-8G Crusader
1955

Tracking and Data Relay Satellite
1983

Manned Maneuvering Unit (MMU)
1984

Mars Pathfinder Lander Prototype
1996

Although discussion continued on a possible annex for the National Air and Space collections during the 1970s, Virginia's voices were, for the most part, silenced when construction began on the National Mall. The museum's staff was buried in the details of creating exhibits, preparing artifacts for installation and display, and hiring new staff as the project moved toward completion. The tasks were endless.

Yet less than 18 months after the museum opened its doors, planning for a long-term solution to artifact storage, collections management, and acquisition policy was officially under way once again. Enter World War II P-40 fighter ace Don "Lope" Lopez, head of the Aeronautics Department. In a November 1977 memo to Director Michael Collins, Lopez suggested that massive buildings, much larger than the exhibit spaces in the new museum, be procured at one of the regional airport facilities. He recognized that modern aircraft were generally larger, heavier, and more difficult to move by road from one location to another. It was essential that future acquisitions, perhaps as large as a Boeing 747, have covered display space and get to travel there without being disassembled, trucked over miles of highway, and then reassembled. Such costs would be prohibitive except in the rarest of cases. Lopez ruled out any location more than one hour's drive from the museum. This criterion narrowed the possibilities significantly.

In 1980, Lopez and Deputy Director Mel Zisfein initiated a study to select the best candidate for the annex. Although Dulles Airport seemed most logical, it took a full | to page 338 |

The arrival of the space shuttle *Enterprise* at Dulles Airport in November 1985 assured that the concept of an annex to the Air and Space Museum would become a reality—but not for another two decades.

PAGES 326–27: On December 15, 2003, the Steven F. Udvar-Hazy Center opened its doors to the public for the first time. Thousands braved long lines during a driving cold rain to be among the first visitors to experience the open, hangarlike environment.

NASA
905

When the Air Force retired most of its SR-71 Blackbirds in the early 1990s, this one (No. 972) was transferred to NASM for eventual display at the Udvar-Hazy Center. Today the Blackbird rests center stage on the museum floor, offering a stunning image from the second-level-entry overlook.

MIC-240
ABM

| from page 334 | three years before the Smithsonian Board of Regents endorsed the plan, and the Federal Aviation Administration offered 100 acres of land to the south of the airport for the construction of the Dulles center. The hope was that the annex would be completed by the turn of the next decade. As it turned out, Lopez's hopes were but a pipe dream. Just like the downtown building, more than two decades would pass between the initial plans and the opening of the facility.

The Steven F. Udvar-Hazy Center felt more like a reality and less a dream as actual planning and design work began. This 1996 architectural drawing of the site plan for the Hazy Center became tangible proof of that reality.

LAYING THE GROUNDWORK

In February 1981, four men from outside of the Smithsonian system—the "gang of four"—began a collaboration that lasted five years and became essential to the eventual construction of the annex. These Dulles region supporters were Carrington Williams, Steven Gelban, Leo J. Schefer, and Thomas G. Moore. This ad hoc group served a vital role in providing access to participants and decision-makers and also lent big-project experience to the process. Start-up was slow, and, as with many projects near the Capital Beltway, big-name support would be needed for success. Because it had been less than a decade since the museum had opened its mall facility, some Castle representatives also felt it was premature. The message concerning the addition of an annex was that "it was not their turn" for yet another.

During 1983, the Smithsonian initiated a number of concurrent studies of feasibility and requirements, while Dulles was updating its airport master plan document. Nine parcels of land on the Dulles property were evaluated, and the two organizations finally chose a plot that met most of their requirements. Out of the gang of four, an informal committee, would come the Air and Space Heritage Council. This group worked tirelessly behind the scenes with museum staff, local politicians, federal officials, and businesses to push forward the idea until it became a reality—a mostly privately funded reality.

SMOKE JUMPER'S HELMET
Ca 2004
Worn by U.S. Forest Service firefighters who descend by parachute

The gang of four hosted a major steering event in 1984 attended by several political and aviation luminaries. Virginia Senator John Warner, former Senator Barry Goldwater, Norman Augustine, Jennings Randolph, and even James H. "Jimmy" Doolittle were there. Discussion that evening centered upon how to gather support for the annex. It was during this dinner that Jimmy Doolittle was heard to remark that, to make it a reality, the museum needed to "nail its colors to the mast" and push full speed ahead. Despite the nautical tone of his comments, his message was clear—press on!

But by 1985 it became apparent that Congress had no interest in passing legislation authorizing the project to move forward—it was a dead end. Or was it? As things turned out, 1985 was truly the pivotal year for the annex. In March, NASA notified the museum that the space shuttle *Enterprise,* the glide test vehicle for the shuttle program, would be made available to the museum. Director Walter Boyne was hesitant to exercise the option to collect the artifact under the NASA–NASM Agreement, which essentially gave the museum the right of first refusal. Where could it go? How would it be protected? Where would it be displayed to the public? During a meeting with Acting FAA Administrator Sandy Murdoch, Boyne raised the subject of the shuttle acquisition. Murdoch immediately advised him to acquire the behemoth vehicle, and, largely based upon that advice, the museum committed to acquiring the shuttle.

On November 18, 1985, NASA's Boeing 747 carrier aircraft delivered the shuttle to Dulles. Public attention was heaped on the arrival when the carrier with the shuttle on its back took an aerial lap around the Beltway before landing at the airport, causing several car accidents. For three weeks, NASA staff worked to unload the massive craft from the carrier and then tow it to its outdoor storage location.

Shuttle delivery was not free, nor was the transport donated by NASA. Senator Warner skillfully reapportioned government funds to cover the costs of the delivery—$1.5 million. The delivery was also useful to NASA, because Dulles served as an emergency landing field for the 747 when carrying the shuttle from coast to coast. No landing tests had ever been done there, so the delivery served that purpose.

That year, the Wright Memorial Trophy Dinner, organized by the Aero Club of Washington each December, was emceed by actor William Shatner, who played Capt. James T. Kirk—captain of the starship *Enterprise*—on the original *Star Trek* series. The space shuttle *Enterprise* became the orphan child around which the Udvar-Hazy Center was eventually | to page 342 |

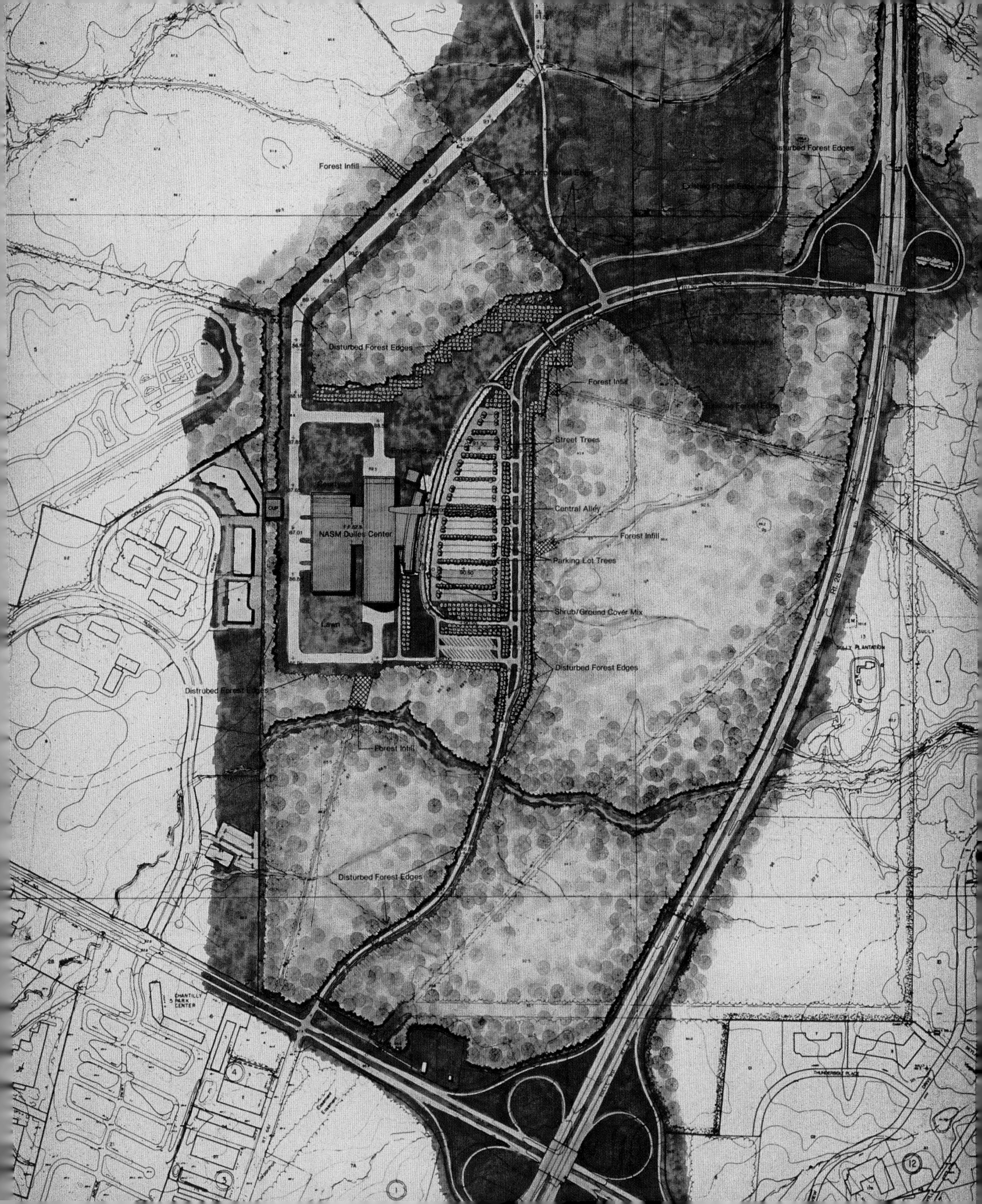

Forest Infill
Disturbed Forest Edges
Disturbed Forest Edges
Forest Infill
Street Trees
Central Alley
NASM Dulles Center
Forest Infill
Parking Lot Trees
Shrub/Ground Cover Mix
Lawn
Disturbed Forest Edges
Distrubed Forest Edges
Forest Infill
Disturbed Forest Edges
RT. 28
SULLY PLANTATION
CHANTILLY PARK CENTER
THUNDERBOLT PLACE

In 1984 Don Engen, then administrator of the Federal Aviation Administration, helped set aside land at Dulles Airport for the museum. Later he became its director and contributed many ideas to the design of the Udvar-Hazy Center, which features an observation tower named for him. Engen wrote these words while serving as the museum's director.

VICE ADM. DONALD ENGEN

ON THE MUSEUM

Early every morning, as I enter the National Air and Space Museum's empty and quiet great halls, which will soon be filled with thousands of visitors, I savor the cool air and walk among the many aviation and space exhibits in awe. Many of the airplanes I have flown. I have not traveled in space yet—other than for those fleeting moments in the Phantom II when I failed to escape the fabled surly bonds of Earth. I can identify with so much of our 94 years of powered flight that when I stand silent and alone in those exhibit halls before the awesome presence of aviation history, I hear ghosts. Many, many designers, engineers, pilots, and other dreamers, who preceded me, made possible what I chose to do. . . . I marvel at what pilots and astronauts will know and do in air and space during the years to come.

NASM DULLES CENTER

| from page 338 | built. But it spent a very difficult winter exposed to the elements. At one point in early 1986, snow and ice covered it, initiating the deterioration of many of the plywood parts that were used to form the shape of the engines and maneuvering pods. The *Enterprise* needed protection, pure and simple.

"A unique band of Dulles area citizens," as Lin Ezell (eventually the project coordinator for the annex) called them, joined forces to provide a privately funded home for the *Enterprise*. The ASH Council—established by members of the Washington Dulles Task Force, the Aero Club of Washington, and the National Aviation Club—was chartered to provide private funds to display the shuttle on Dulles property. All of these groups were capable of reaching out to powerful political forces as well as potential donors that might contribute to making the annex a reality.

Their first step was to provide funding for the completion of the Smithsonian's 1983 feasibility study. Don Lopez, now deputy director, chaired the nine-member task force from a cross-section of the museum staff, whose task was to provide a "needs list" for the new facility. The company of Dewberry & Davis Engineers, Architects, Planners, and Surveyors wrote the report. The southernmost Dulles site was selected by both teams as the most desirable for construction, access, and airport noninterference. Specifications for climate control and weight allowances were included in planning documents. Then pencils hit the paper and concept drawings were prepared by Dewberry & Davis. This marked the first concrete steps toward | to page 347 |

William "Jake" Jacobs, NASM chief exhibits designer for the Udvar-Hazy Center, holds a Plexiglas aircraft cutout above a floor plan for the center's main exhibit hangar during a planning exercise. By the time the center opened, the entire layout had been computerized, making physical and paper plans obsolete.

MONOCOUPE 70
NOMAD
BEECH 35
BONANZA
SGU 2-22
EXTRA
260
PITTS
S-1C

The main supports for the aviation hangar were assembled from the ground first and then the top arch was lowered into place, bolted, and then welded. The exterior roofing and coating was installed in large sections as well, from north to south.

| from page 342 | the construction of the annex, but 18 more years would pass before a single visitor would be able to marvel at the building and its precious contents. While the National Air and Space Museum staff clearly recognized the need for expanded storage and display space for their growing collection of planes and space artifacts, others in the Smithsonian did not.

In 1988, the museum scraped together funds for a temporary building to protect the shuttle, not from any capital building account, but rather from monies put away and traditionally used to fund an academic fellowship in Aeronautics. The Shuttle Hangar, a steel-framed cube, was located within the Dulles secure access zone and eventually housed many more artifacts than just the shuttle. It is still used for object storage in 2010. But in the larger plans for the annex, political and bureaucratic obstacles were encountered at every turn, as would be expected in such an ambitious project.

Also in 1988, the Smithsonian Office of Design and Construction hired Skidmore, Owings & Merrill (SOM) to produce a formal planning study that would provide a level of detail in concept and design not yet attempted in the process. After four months, the study was complete. It was this SOM report that established the "one museum, two locations" concept for the Dulles center's role in the museum. Despite protest from some of his senior staff, Smithsonian Secretary Robert McC. Adams recognized that, despite a few bureaucratic imperfections in the document, the museum was remarkably popular with the visiting public because its exhibits portrayed achievements that are "unambiguously ours." So he took the report to the Smithsonian Regents and asked for permission to lease 200 acres of property at the airport and $300 million to build the facility. Meanwhile, efforts were made by the ASH Council to gain funding from Virginia for infrastructure that would eventually allow commercial interests to expand into the Dulles corridor after the facility was opened. Discussions also included suggestions that Metrorail expansion to Dulles should include a stop at the center to facilitate an easy method to travel to the facility anywhere within the greater Washington area.

Recognizing that approval for the construction of the annex was no longer an "if" but a "when," airport facilities from across the nation offered proposals to the Smithsonian under the guise of federal cost savings and fairness. Congressional authorities, particularly the representative from the Denver Airport region, David Skaggs, insisted that all offers, particularly the one from his district, be considered by Air and Space. These efforts were mostly last-ditch efforts that were not well thought through. This evaluation process, however, resulted in more delays, and the 1990 legislation introduced in both the House and the Senate died yet again. But once all the alternative proposals were evaluated between December 1990 and early 1993 and discarded, the fight over the annex's permanent location was over for good.

As had been the case with the construction of the downtown facility, these delays inevitably increased costs. The new National Air and Space Museum director since 1987, Martin Harwit, was forced to establish hard limits for costs and character of the new building. These restrictions were established only after the arrival at Dulles International Airport of one of the most significant artifacts in the collection—the Lockheed SR-71 Blackbird.

On a cold, overcast day in early March 1990, a large crowd gathered on the observation deck at Dulles. Numerous officials from the Pentagon and museum staff awaited the arrival of the museum's newest acquisition. This was not a normal delivery. The tower had cleared commercial air traffic around the airport so as to make the airspace available for the landing. After nearly 30 years of service, the Air Force was retiring its Blackbirds. For many in the crowd, this would be their first and last opportunity to see a flying SR-71. In the distance to the west, the first view was its bright landing light. On this gray day, the aircraft's slim profile made it difficult to see. Lt. Col. Ed Yeilding and his radar systems operator, Lt. Col. Joseph Vida, had just set the blistering coast-to-coast speed record of 1 hour, 4 minutes, and 20 seconds, averaging a speed of 2,124 miles an hour. Soon the airplane made a low and slow pass in front of the crowd.

The expectation was that the Blackbird would land, but to the surprise of the gathering, Yeilding turned out of the landing pattern and began a second pass. Just in front of the podium he put the aircraft into full afterburner. With a terrific roar, a hot blue flame in a distinctive diamond-patterned shock wave equal in length to the aircraft shot out from the tail. Apparently, this was the crew's comment on the wisdom of retiring the SR-71. | to page 354 |

BELL ROCKET BELT
1960s
Also called a jet pack; uses noncombustible gas as a propellant

OPPOSITE, TOP: During April 2002, Steven and Christine Udvar-Hazy tour the construction of the center that today bears his name. Hazy amassed his fortune leasing aircraft to the airline industry and donated more than $60 million for the construction.

OPPOSITE, BOTTOM: Only 14 months before opening, the center had no concrete floor and the interior had not been painted its light-reflective white. A mild winter helped contractors complete the main hangar interior ahead of schedule and under budget.

Construction proceeded from north to south (right to left in this photograph). Once the arches were up, the special coated roofing followed, much like dominoes in reverse, while construction on the IMAX Theater and the Engen Tower proceeded concurrently. Virginia red clay became a part of everyday life.

The only remaining Seiran was designed by the Japanese to attack New York City and Washington, D.C. By war's end, plans had changed. After a scrubbed plan to attack the Panama Canal, the target shifted to naval vessels moored at Ulithi Atoll waiting to invade mainland Japan in 1945.

The M6A1 Seiran was designed to operate from a submarine. The main wing spar rotated 90 degrees and then folded back flat against the fuselage. Two-thirds of each side of the horizontal stabilizer and the tip of the vertical stabilizer also folded down.

AICHI M6A1 SEIRAN

WORLD'S LARGEST SUBMARINE-CARRIED AIRCRAFT

For Japanese war planners, the Seiran was a means to attack the American mainland. To support Seiran operations, the Japanese developed a fleet of gigantic submarine aircraft carriers that could travel 43,000 miles carrying three Seirans in a waterproof hangar.

Original attack plans included strikes on New York City and Washington, D.C. Those ideas were soon abandoned as Allied forces flooded into the Pacific via the Panama Canal. Japanese planners chose to strike the canal with ten Seirans loaded with six torpedoes and four bombs. Before it was launched, the Japanese decided instead to strike the invasion fleet anchored at Ulithi Atoll with six Seirans and four Nakajima C6N1 MYRT reconnaissance aircraft. The war ended before the submarines surfaced to attack.

The museum's M6A1 is the only surviving example of a Seiran. Shipped

Seen in its handling cradle, the Seiran was to be launched without its floats from a submarine designed to house three aircraft. OPPOSITE, TOP: Two I-400 class subs were loaded with six Seirans in August 1945 to attack Ulithi Atoll but were submerged when Emperor Hirohito announced Japan's surrender. OPPOSITE, BOTTOM: Folded in storage mode, three aircraft could be loaded into this watertight hangar deck on the I-400 class submarine.

to the U.S. and eventually transferred by the Navy to the museum, it arrived at the Garber Facility in November 1962 but remained outdoors for the next 12 years. Deterioration of the plane was severe by the time restoration began in 1989.

No production drawings survived. A team of staff experts, volunteers, and Japanese nationals conducted exhaustive research into how the systems operated, and then they accurately reconstructed missing components.

A metal flap was damaged and hastily covered with fabric patches. Fuel tank interiors were contaminated with paper documents, and parts fit together poorly. Inside one wing panel, the complete English-language alphabet was scratched into the metal.

The project took more than a decade to complete, and its success is a testament to the remarkable skills of the Garber restoration staff.

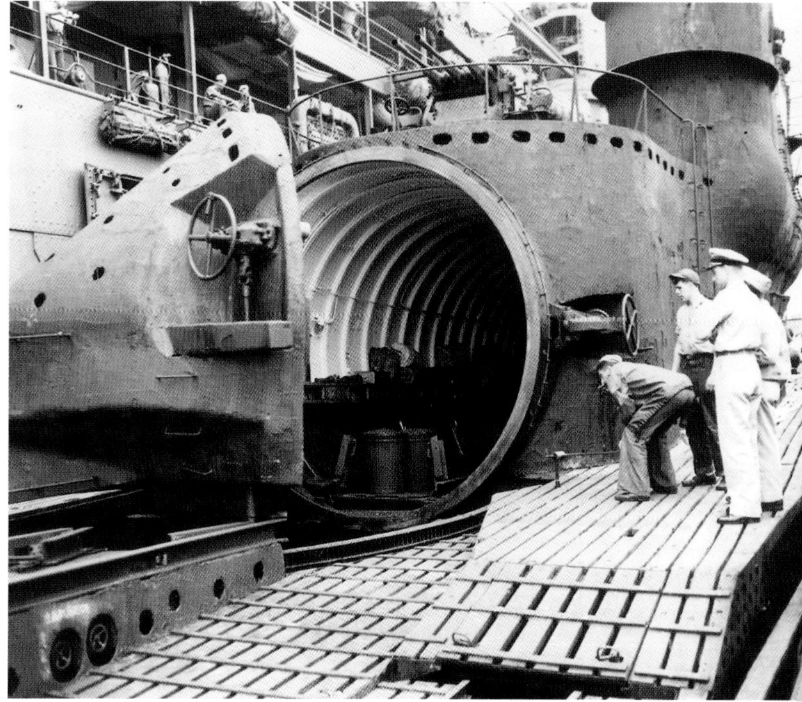

| from page 347 | Only after a spectacular pull-up maneuver did they come around again for a landing.

Seven months later, Harwit directed that the total cost for the center remain below $330 million and the initial construction (Phase One) be capped at $162 million. In essence, this dictum set a time clock in motion. For each passing year that funding was not made available, the building would shrink and the amenities within the building would become more limited. Immediately, the original 1.5-million-square-foot design was cut by more than half to a still massive but disappointing 670,000 square feet. Later, the design stabilized at 760,000 square feet, including the Boeing Aviation Hangar, the James S. McDonnell Space Hangar, and the Mary Baker Engen Restoration Facility.

Finally, in February 1993, Representatives Norman Y. Mineta, Joseph M. McDade, and William H. Natcher introduced House bill H.R. 847. In the Senate, John Warner, Daniel P. Moynihan, and James R. Sasser introduced S. 535. Both bills authorized the Smithsonian to plan and design the Dulles extension and included eight million federal dollars for that purpose. President Bill Clinton signed the final bill into law (Public Law 103-57) on August 2, 1993. That was the end of federal appropriations for the center.

The idea of including a parking fee for the annex's visitors actually originated in 1985. The idea was to deter "squatting" on museum property that might come from Dulles airport users, thus undercutting airport parking revenue. Additionally, since the facility was going to be built with private funds, private business practices for long-term fund-raising were incorporated in early plans to house the shuttle at the airport. Contract-run paid parking and an IMAX movie theater at the center would provide a revenue stream to repay building debt. Now bureaucratic momentum began to build. Federal dollars had been given to put pencil to paper and once that happened, the only thing left to do was raise the requisite $162 million to begin Phase One.

OPPOSITE: This view from the entryway looking out toward the parking lot highlights the Wall of Honor and the sculpture "Ascent" by John Safer. The delicate upward spiral represents the climbing flight path of an airplane on initial takeoff.

LINDBERGH MEDALLION Ca 1927 Depicts the transatlantic flight route from New York to Paris.

The political fallout of the 1994–95 battle over the *Enola Gay* exhibition resulted in Harwit's departure and the 1996 arrival in the director's chair of a former head of the Federal Aviation Administration, Vice Adm. Donald D. Engen. He was familiar with Washington politics and was well respected nationally in the aerospace industry. His job was to get the center built. In 1984, while FAA administrator, he had been instrumental in setting aside the land at Dulles Airport. Soon after becoming the director, Engen traveled across the country to meet with aviation enthusiasts, telling them about the need for the center and soliciting their support. The new facility became his legacy, as Engen unfortunately was killed in a glider accident three years later in July 1999.

From 1995 through 1999, fund-raising for the annex had been tough. Only $28 million of the requisite $130 million to begin work had been raised. Then, late in the summer of 1999, Don Lopez, now acting director, gave a tour of the Garber Restoration Facility to a young entrepreneur, Steven F. Udvar-Hazy. The Hungarian immigrant had made his fortune by leasing airliners. In October, Udvar-Hazy committed to a $60 million personal contribution to the annex. Suddenly there was new hope that it might soon become a reality. Udvar-Hazy later added another five million dollars to his contribution. On March 28, 2000, the Smithsonian named the facility the Steven F. Udvar-Hazy Center in recognition of his philanthropy.

To fill the directorship, similar political and fund-raising skills seemed a prerequisite. The search committee soon fixed on retired Marine Corps general and top NASA official John R. "Jack" Dailey. He walked in to his museum office at the beginning of January 2000, and from that day forward, one of his primary jobs has been fund-raising for the center.

Dailey added to the momentum that Engen had built by increasing the size of the facility and establishing an ambitious opening date, during the national Centennial of Flight celebration, which culminated in the hundredth anniversary of the Wright brothers' Kitty Hawk flights of December 17, 1903. Ground needed to be broken by early 2001 or there would be no way to complete the project in time. The only way to do that would be to start construction before the full amount was raised. Roads would need to be built, acreage cleared, steel formed, and plans finalized. There was no time to lose.

Sheila Burke, then Dailey's superior as Smithsonian undersecretary for American Museums and National Programs, agreed with the museum's proposal to adopt a "phased approach" to center construction. She reasoned,

The first aircraft installed at the Udvar-Hazy Center was the venerable Piper Cub. Flown by hundreds of thousands of pilots, the Cub represents the pervasiveness of aviation in America during the past half century. The Cub is dwarfed by the massive size of the center.

like General Dailey, that initiating the contract fixed the price, established future fund-raising deadlines, permitted calculation of operating costs, offered "unique opportunities to entice potential donors," and limited the Smithsonian's financial risk. The Smithsonian Board of Regents agreed with Burke on March 28, 2001. The next day, following an earlier contractor selection process, Hensel Phelps Construction Company was awarded the construction contract for the building. The race was on to beat the December 2003 opening date.

Assisted by an unseasonably mild winter in 2001, construction fortunately ran a bit ahead of schedule. Once the "roof was on the barn," everyone began to breathe a little easier. The interior could be completed without competition from the elements. Project coordinator Linda "Lin" Ezell expertly handled complicated schedules and multiple contractors with aplomb. After the center opened, her final task was to assemble images and records for *Building America's Hangar: The Design and Construction of the Steven F. Udvar-Hazy Center,* which was published in time for the public opening of the Space Hangar in fall 2004.

William "Jake" Jacobs, an experienced exhibition designer at the museum, led the team that created the plans for artifact displays throughout the museum. By using computer-assisted design, Jacobs was able to "build" the entire museum in virtual reality. This task included making sure objects did not conflict with structures and determining the weight limits and cable lengths for aircraft and artifact suspension, as well as | *to page 361* |

From 1974 to 1984 Lin Ezell was a historian at NASA, where she authored or co-authored several books. In 1984 she joined the National Air and Space Museum staff as a curator, then rose through the ranks to become assistant director for collections management. In that position she guided the Garber Facility for a decade. But Lin made her most dramatic impact when she was selected as project coordinator for the design and construction of the Steven F. Udvar-Hazy Center.

LINDA NEUMAN EZELL

ON THE STEVEN F. UDVAR-HAZY CENTER

At museums we come face to face with real examples of what we heard about as children—dinosaurs or steam locomotives—or what we actually used in our everyday life—Flash Gordon lunchboxes or kitchen gadgets. Or we finally see for ourselves objects made familiar only through books and television—the spacecraft that brought home the first men from their voyage to the Moon. Or we behold a symbol long held sacred by a free people—the Star Spangled Banner. We have a basic need to see for ourselves these special icons of history, these common reminders of our personal history. Museums help make this happen. . . .

We opened the new museum at Dulles on December 15, 2003. During the first six months, over 1,000,000 visitors found their way to our door. They came to stand next to wood and fabric aircraft that had flown when the age of powered flight was new, when aircraft flew over European skies angry with war. Icons of World War II, from all theaters and most of the combatant nations, were a draw. The museum finally had enough room at the Udvar-Hazy Center to assemble and exhibit Enola Gay, *the B-29 that dropped the first atomic weapon, which led to the final surrender of that global war. Large aircraft, like the Concorde and the first Boeing 707, graced the hangar floor, while a legion of planes seemingly flew overhead in the mammoth space. And would-be astronauts came face to face with a Space Shuttle. . . .*

The center was designed with the expectation that three to four million visitors would frequent the facility annually, and we are on our way.

The Udvar-Hazy Center . . . is a photogenic building. We were blessed with several talented photographers who regularly recorded our progress, from the last old farm fence post coming down to the crowds arriving on opening day. They shot bare steel going up on blustery winter days, and they slogged through red mud when I asked them to record yet another construction milestone. They were up on lifts, rooftops, and ladders. They captured the morning sun's reflection on glass and metal skin and the moon's grace against stainless steel sculpture. And they didn't forget the people who made it all happen.

It has been an honor to be part of the team that opened the doors of a truly magnificent facility. . . . Every process was important, every person left their mark. . . . I think we've made a difference.

Smithson
National
Steven F.

A reproduction of the Wright Flyer "soared" above the crowd who gathered for the formal opening ceremonies at the Udvar-Hazy Center.

| *from page 357* | the locations for the educational exhibit stations being created by the museum curatorial staff.

FILLING THE CENTER

On March 17, 2003, the Piper Cub became the first airplane delivered to the Udvar-Hazy Center from Garber. Perhaps this was fitting, as more men and women had learned to fly in this aircraft than any other trainer ever built. After it had been assembled near the museum catwalk in preparation for its eventual suspension from the arcing rafters, the image was both significant and comical. This tiny yellow plane sat "alone, unarmed, and unafraid" in the middle of the floor. It was the first of what would eventually total 80 aircraft before opening day. On the last day of March, the Garber Facility closed to the public.

During the next six months, dedicated professionals from the Garber Facility spent every day preparing, loading, moving, unloading, and assembling those aircraft that had been selected for display at the center. From before dawn until after dusk trucks loaded with those craft traversed the Beltway to deliver their precious cargo to the center. Some aircraft, like the Air France Concorde, were flown directly to the center. The Concorde landed at Dulles on steamy July 12 after a flight from Paris in preparation for installation at the center. Among the passengers on its last supersonic, transatlantic flight was Bob van der Linden, the curator of the aircraft.

The Boeing 307 Stratoliner, the pioneering, four-engine, propeller-driven airliner, arrived at Dulles on

August 6. The Stratoliner had been front-page news when the plane had to ditch in shallow waters while practicing approaches at Boeing Field outside Seattle on May 28, 2002. The company subsequently restored it to pristine condition. The overall investment made by that company to put the 307 into flying condition twice, and to restore the groundbreaking prototype of the 707 airliner, the 367-80, which also came to the center, was significant.

By far the most time consuming and tedious reassembly project was that of yet another Boeing aircraft, the B-29 Superfortress *Enola Gay*. For nearly five months, a team of six (sometimes more) took the component parts and rebuilt the plane—completing a project that had taken more than two decades and totaled more than 300,000 working hours. The restoration and reassembly of the aircraft was the single most all encompassing restoration project ever accomplished by the museum staff—by a long shot. It continues even today, as curators are still on the lookout for those few remaining parts that will make the aircraft 100 percent complete.

OPENING DAY

The weather was terrible during the Wright centennial week. A cold rain had succeeded in ruining the attempted re-creation of the first flight at Kitty Hawk, North Carolina, by the Wright Experience, a group of replica specialists from Warrenton, Virginia. On the morning of Monday, December 15, 2003, throngs of visitors lined up outside of the Udvar-Hazy Center in anticipation of the vistas and history soon to be revealed to the public. Pre-opening events had garnered a huge amount of local publicity. One event in particular, the military veterans' event held on December 9, had included more than 4,000 attendees; the reviews were nothing less than magnificent. Museum Special Events director, Linda Hicks, recalled that the Salute to Veterans—the first big event to mark the opening of the Udvar-Hazy Center—was created "to fulfill a promise to Veterans—in particular WWII Veterans—that when the *Enola Gay* was totally put together NASM would invite our Nation's veterans to be the first to celebrate in the bomber's new home."

OPPOSITE, TOP: Restoration can take thousands of hours. Here the Kawanishi N1K2-J Shiden Kai (GEORGE) undergoes assembly at the center in 2004.

OPPOSITE, BOTTOM: Not designed to be disassembled, the B-29 posed interesting problems during restoration, which took more than 300,000 man-hours.

DOLL WEARING SPACESUIT Ca 1962
Fewer than ten were manufactured by the B. F. Goodrich Company to be given as gifts to visiting dignitaries.

But the event became much more than just a celebration. As General Dailey stated in his invitation letter, "The salute will honor veterans of all eras, including active duty military." It was a joyfully patriotic day for all who attended. That late afternoon event featured music by the U.S. Air Force's World War II–style jazz band, Airmen of Note, and remarks by the Chairman of the Joint Chiefs of Staff, Gen. Richard B. Myers.

The opening gala was held during the evening of December 10 at the center, followed by the official dedication on the morning of December 11—attended by the Vice President, politicians, aviation legends, Smithsonian staff, and local, national, and international media. Guests were met at the entrance to the center by a staff member who escorted them to their carefully assigned seats at the far south end of the Aviation Hangar. The journey took the visitors through a tunnel of black cloth until they emerged into a seating area in front of a large stage, a huge television screen, and dramatic lighting. A replica of the Wright Flyer was suspended some 30 feet above the gathering crowd. A massive shroud of black cloth was suspended behind the stage, preventing any view of the interior of the center.

The chancellor of the Smithsonian Institution, Chief Justice William H. Rehnquist, accepted the Udvar-Hazy Center on behalf of the Board of Regents. As Chief Justice Rehnquist concluded his remarks, a fanfare was played; the replica Flyer actually "flew" to a landing zone near the stage from its perch behind and above the crowd; and the dark shroud fell to the floor, revealing the interior of the new museum. The Udvar-Hazy Center, built with private and corporate donations and some $40 million in infrastructure support and construction by the Commonwealth of Virginia, now belonged to the Smithsonian Institution.

The next Monday, opening day, was extremely cold, but the parking lot was full and the lines were long. Museum security was on full alert and included not only uniformed guards but also a sizable contingent of plainclothes officers. Museum staff had been carefully monitoring several antinuclear organizations that had planned to stage public protests near the *Enola Gay*. A Plexiglas barrier had even been erected along the catwalk to protect the aircraft from direct assaults.

Dailey and the museum leadership knew there would be antinuclear protests near the *Enola Gay* that day. In fact, the action had been carefully choreographed between the protesting groups and the museum. On the ground level, a few dozen demonstrators unfurled pieces of cloth that could be combined to spell out antinuclear phrases. With the foreknowledge of museum officials, the protest organizers had notified the local media of

A
343-35

ENOLA
GAY
82
NO SMOKING WITHIN
100 FT.
HYSTER

their intentions, and several stations had planned to cover the events of the day anyway. The activists were well behaved, singing songs, and waving their banners and papers in front of any TV camera lens they could find. Unfortunately, a man unaffiliated with the protestors hurled a small glass jar of red paint at the aircraft's nose from around the barrier on the catwalk above. It caused a small dent in the airplane and shattered when it fell to the concrete floor, necessitating a cleanup. Protest groups were politely escorted to the main entrance, where most departed to meet their buses. Other than this incident, which received minimal media coverage, opening day was an enjoyable and well-attended event.

In the first two weeks, the center welcomed more than 200,000 visitors, and an additional half million came in the next three months. On June 9, 2004, the one millionth visitor passed through the doors at the Udvar-Hazy Center. On display were 80 aircraft; in addition, the Mercury 15B and Gemini VII capsules and the space shuttle *Enterprise* could be viewed from the entryway to the James S. McDonnell Space Hangar. The hangar was named for the founder of McDonnell Aircraft, the St. Louis company that had built the Mercury and Gemini spacecraft. The rest of that hangar would be filled, mostly during the next year. Also available to visitors were the IMAX movie theater (later renamed the Airbus IMAX Theater), the gift shop, the cafeteria, and the Donald D. Engen Tower, which overlooks Dulles Airport and has exhibits on air traffic control and airport operations.

The Boeing Aviation Hangar, the larger of the two exhibit areas in the center at just over 235,000 square feet,

TOP: After arrival in 1985, the shuttle *Enterprise* remained outdoors for months. The elements had taken their toll, and by 2004 restoration teams had much challenging work to do. Here, restoration specialists paint the payload bay doors.

BOTTOM: Staff members of the Collections Division carefully push the P-61 Black Widow into position at the Udvar-Hazy Center.

was named in honor of the Boeing Company in 2006, whose generous support of the Udvar-Hazy Center has helped to complete the new restoration facility and archives. It features aircraft hanging at several levels, suspended from the building's huge trusses, and larger, heavier aircraft displayed on the hangar floor. The suspended aircraft are displayed at various angles to demonstrate typical flight maneuvers—an aerobatic airplane suspended upside down, a World War II fighter angling for a victory, and a small two-seater flying level. Multi-level walkways rising about four stories above the floor provide a wide variety of angles from which to view the aircraft and spacecraft on display. At the Udvar-Hazy Center, the ability to view these aircraft and spacecraft in three dimensions is made possible by moving the visitor around the objects at different levels. The view from the upper catwalk is particularly breathtaking.

Aerobatic, general aviation, commercial, and World War I and II aircraft are located to the south of the second-level-entry overlook. To the north, visitors encounter the post–World War II military aircraft collection, including Russian MiGs, carrier aircraft, military helicopters, and land-based fighters. One of the most memorable vistas, the entryway overlook, provides an opportunity to view the sleek Lockheed SR-71 Blackbird, the fastest airplane ever built, from above and from the front—much like it would be seen during midair refueling. Beyond the Blackbird can be seen the shuttle *Enterprise.*

After the opening, the Garber restoration staff's mission took a decided turn for the next 18 months from air to space. Restorers went to work at Silver Hill to fill the McDonnell Space Hangar with artifacts, even as *Enterprise* restoration was ongoing inside the hangar. Opened November 1, 2004, the 53,000-square-foot hall features hundreds of artifacts arrayed around the *Enterprise*. Objects range from the 69-foot floor-to-ceiling Redstone missile to tiny "Anita," a spider carried on Skylab for web-formation experiments. The hangar and its holdings illustrate the scope of the Smithsonian's rocketry and space collections as organized around four main themes: rockets and missiles, human spaceflight, application satellites, and space science.

More than 110 large artifacts are housed in the hangar. The biggest and heaviest—including the F-1 main engine for the Saturn V rocket and a space shuttle main engine—are displayed at ground level, except for the 22-foot-diameter Saturn V instrument ring, which is mounted on stilts. An array of cruise missiles, satellites, and space telescopes hangs above. Two elevated overlooks allow visitors to study suspended artifacts straight on and ground-level displays from above. In 2010, these balconies were connected by a mezzanine that now allows visitors to look down, on one side, through glass into the new restoration workshop constructed during Phase Two of the center, and on the other side, through an open overlook into the Space Hangar. More than 500 smaller artifacts are exhibited in customized cases throughout the hangar, including advanced space-suit prototypes, research crystals formed in orbit, sounding rocket payloads, space-themed toys from the 1950s and 1960s, and even borscht in tubes, prepared for Soviet cosmonauts.

SPACESHIPONE GLOVES
2004
Worn by test pilot Mike Melvill on the first privately launched space missions

EXHIBITION PHILOSOPHY

The original National Air and Space Museum building was constructed of a collection of cubical gallery spaces, each designed to tell one specific aviation or space story. When completed, each consisted of a mix of traditional museum exhibit text (the story) that supported a number of artifacts (the technology) on display in each of the first- and second-floor spaces. Yet only a small fraction of the nation's air and space artifact collection was displayed. The single largest aircraft in the museum was a 1930s vintage Douglas DC-3 with a wingspan just under 100 feet across.

The essence of the museum's story line was to celebrate American ingenuity while memorializing the nation's role in the development of aviation and space technology around the world. Not much space was devoted to examination of the impact of aviation or spaceflight on the human condition or the consequences of this technology upon the economy, the conduct of war, or the politics of aerospace technology at home or abroad. For the most part, the curators assumed that aviation and space science and technology was inherently a "good" thing—a source of great pride for the country. From the very beginning, the National Air and Space Museum was a triumphant experience for the many millions of Americans who waited, sometimes for hours on end, to enter the sanctuary and marvel at these vehicles of flight, and millions of international visitors experienced it much the same way.

While museum curators have updated galleries | to page 368 |

The Northrop P-61 Black Widow highlights the skill of NASM's restoration staff. During preparation for display, four layers of paint were carefully removed one by one, revealing the markings of the airplane's service history in the Army Air Forces, the NACA, and the U.S. Air Force.

| from page 365 | and pushed forward into more contextually rich storytelling, with success greater in some cases than others, plans for the Udvar-Hazy Center did not provide an exhibit environment anything like that created downtown. The center was designed to hold as many artifacts as it possibly could, with little space for panels containing detailed stories related to the history of the artifacts or of aviation and spaceflight in America and the world. In essence, the center would be a massive open storage area that could house the collection while permitting visitors to view the artifacts fully assembled. The pressures of fund-raising, however, which made a storage facility difficult to sell, pushed museum directors and curators toward making it into a second museum. The IMAX theater, a cafeteria, and a gift shop were added for revenue generation, and at a late stage, the museum decided to insert exhibit stations to give at least some context for groupings of artifacts, like commercial airliners, Korean and Vietnam War military aircraft, or human spacecraft.

At the center, many of the artifacts' cockpits were photographed using a technique called QTVR 360-degree imagery. A sequence of images was taken with a wide-angle lens using a pattern that provided coverage for the complete interior of the cockpit. That high-definition imagery was then processed with computer software that seamlessly blended the images together in a virtual three-dimensional world. Kiosks containing the computer files for each vehicle were dispersed throughout the center so that visitors would be able to access the images by interacting with a touch screen. Once a visitor selects a particular artifact, by using her finger on the screen, she is able to look up or down (in fact, all around) and zoom in or out to see any object captured during the imaging process. Since physical access to these cockpits in not possible, this technology allows virtual access to objects never possible previously.

OPPOSITE: The Boeing Aviation Hangar will one day hold nearly 80 percent of the national aircraft collection. On opening day, 80 planes were on display. Nearly a decade later, almost 200 aircraft can be viewed each day.

SOUTHWEST AIRLINES PILOT'S JACKET Ca 2000 Bearing the name of longtime airline president, Herb Kelleher

There remains a stark difference between the center's character and that of the original building on the mall, but the two buildings nonetheless complement each other and provide two completely different experiences for the visitor. One provides much more historical, scientific, and technological contexts for understanding aerospace technology; the other allows the visitor to experience the full scope of the best collection of air and space objects owned by any museum in the world.

A NEW RHYTHM

After the McDonnell Space Hangar became accessible to the public, a new rhythm began to build. The process of preparing artifacts for display began to shift back to aircraft. Education programs at the center—similar to those at the mall museum—became the hallmark of daily operations, and special events took on a life of their own.

Although staffing was minimal before the move of collections and archives staff to Phase Two, one division there has been fully staffed and operational from the outset—Education. Aside from permanent museum educators, the center is home to "educators in residence" from surrounding counties. These teachers, selected after a highly competitive process, act as liaisons to their counties to ensure that education programs at the center are fully utilized at all times, which was part of the deal with Virginia to secure the commonwealth's support for roads and other infrastructure for the facility.

Just as is true in the mall museum in recent years, the educators, curators, support staff, and volunteers have worked hard to establish a program of activities designed to attract all ages to the Udvar-Hazy Center. Regular "family days" seek to educate a broad audience in the history, science, and technology of flight. One notable success is "Become a Pilot" Family Day, first held in June 2005, which features a complement of aircraft flown to Dulles and then taxied to the center for static display at the north end of the building. Informational tables are set up inside the building and space is set aside to deliver a short safety briefing prior to exiting to see the aircraft and meet the pilots on the tarmac. Routinely held on the Saturday of Father's Day weekend, "Become a Pilot" Day is one of the three busiest days of the year for attendance. Since 2005, the center has also been the home of Air & Scare—an annual Halloween event for children and their parents.

As more and more artifacts find their way to display space at the Udvar-Hazy Center, interest in filming these rare objects has increased dramatically. Historically oriented programming has seen the most growth. Programs such as *History Detectives* and *Man, Moment, Machine,* | to page 374 |

AIR FRANCE
WT
VMFA-232
7307
D3
MARINES
MODERN MILITARY Aviation

The X-35A was the first experimental model of America's next fighter, the F-35 Lightning II. Powered by a monstrous F-119 afterburning jet engine, the F-35 carries its weapons internally in high-threat combat environments—part of its overall stealth capability.

X-35

AN AERIAL HAT TRICK

The X-35A aircraft—the version designed for the U.S. Air Force—made 27 experimental test flights in 2001. Easily meeting its program goals, the aircraft went back to manufacturer Lockheed Martin for modification into the X-35B, a short-takeoff and vertical-landing version sought by the Marine Corps and the Royal Air Force. The F-35B will replace the obsolescent Harrier Jump Jet and will be deployed on aircraft carriers.

The X-35B's innovative propulsion system received the 2001 Collier Trophy for the greatest aerial accomplishment of the year. A variable exhaust nozzle can rotate downward during flight. Just behind the cockpit, Lockheed Martin implanted a Rolls Royce "lift-fan" into the airframe. A titanium shaft connects it to the engine, and when needed, clamshell doors covering the inlet and exhaust open, and special gears linked to the

Housing a unique, Collier Trophy–winning propulsion system, the X-35B STOVL (short take-off/vertical landing) was the technology demonstrator for the F-35 Lightning II fighter. Both a rotating exhaust nozzle and internal lift fan provide thrust needed for vertical flight.

In the standard configuration, the X-35A operates like a typical fighter plane. This original X-35A was reconfigured to hold the lift-fan system, and then the aircraft was redesignated the X-35B. OPPOSITE: The X-35B's afterburner section reflects the symmetric power and beauty found at the leading edge of propulsion technology.

drive shaft engage to power the massive fan, shooting a blast of air downward. At the same time, thrust vector nozzles deflect engine's jet exhaust downward at the rear. Working together, the two allow vertical takeoffs and landings.

The meaning of the tail markings on the X-35B is a mystery to some. An upside-down top hat is set on a bright red background. From within the hat, three aces are displayed, as if part of a winning poker hand. The lead test pilot for the program explained that the three aces represent the three different types of landings that the program was designed to demonstrate—conventional, vertical, and carrier.

In sports, a hat trick is three individual successes in one game. The radio call sign used by the pilots during test flights reflected the tail marking. Since call signs are restricted to only seven letters, the actual spelling of the name became "HATTRIK."

| from page 368 | and networks such as the Smithsonian's own HD channel, the Military Channel, and National Geographic, have filmed objects and curatorial staff at the center for presentation in their programs, just as they have shot more and more footage in the mall building. When such filming can be permitted, exposure for museum curators and collections is enhanced and, over time, increases visitorship.

In 2008, Hollywood companies filmed major motion pictures at both museum locations. On the National Mall, crews from 20th Century Fox shot scenes for *Night at the Museum: Battle of the Smithsonian,* the first feature-length movie produced with the cooperation of the Institution. Included were fantasy scenes inside the museum building, with Amy Adams as Amelia Earhart piloting the 1903 Wright Flyer and her red Vega. At the Dulles center, Paramount Studios and DreamWorks Pictures filmed sequences with stars Shia LaBeouf, Megan Fox, and John Turturro for *Transformers: Revenge of the Fallen.* IMAX-scale images highlighted the exterior and interior of the building and then concentrated on the SR-71—a robot character named Jetfire. The two movies introduced the museum to a new generation of enthusiastic visitors.

FEATHERING THE NEST

What has been called Phase Two was originally included in the overall Udvar-Hazy construction project. Because raising the entire $312 million was unlikely by 2000, the restoration, archives, and storage building were separated from the construction of Phase One, which included the main hangars, the office area, the theater and the Engen Tower. It was this strategy that convinced Castle authorities that the museum fundraising and construction plans were sound. While fund-raising for the Udvar-Hazy Center never stopped during these years, the two-phase approach allowed the museum development staff to focus their efforts on one project or the other depending upon the donor. Despite the remarkable contributions that the Udvar-Hazy Center has made to the Smithsonian and the National Air and Space Museum, no federal funds have been approved to subsidize the cost of construction of any buildings at the center. The only appropriated funds were the original eight million dollars for the center design study and a smaller amount recently approved to facilitate the move of artifacts and facilities to Phase Two.

The new Udvar-Hazy building is dedicated to the preservation, restoration, and long-term storage and display of many of the most iconic and rare artifacts in aerospace history. From second-level observation areas overlooking Mary Baker Engen Restoration Hangar (named for the wife of the former director after a generous gift from the family), visitors are able to witness some of the magic that occurs behind the scenes, watch museum experts at work on a variety of aircraft and spacecraft, and appreciate the scope and scale of the meticulous labor accomplished before an artifact is placed on permanent display.

The operation is a far cry from what the museum was capable of at the opening of the mall building in 1976.

Space shuttle *Enterprise* is wheeled into the McDonnell Space Hangar on November 20, 2003.

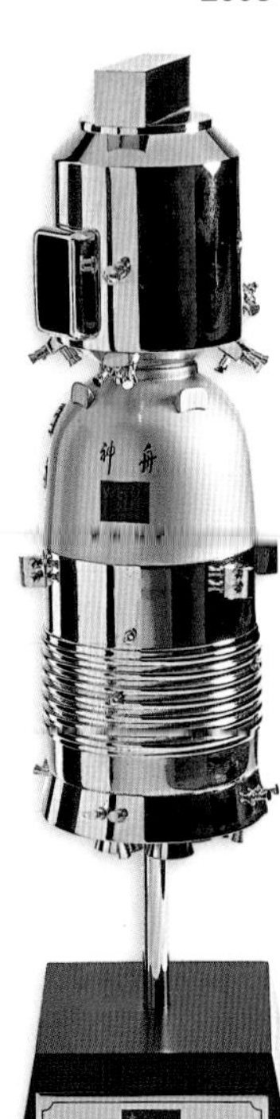

SHENZOU 5 SPACECRAFT MODEL
Ca 2004
China's first astronaut, Yang Liwei, flew in this spacecraft on October 15, 2003

Enterprise
United States

Enterprise

The James S. McDonnell Space Hangar was fully opened to the public in 2004 and today is nearly completely outfitted with artifacts and exhibits.

Professional conservators are very much a part of the team, as they had been in the last two decades at Garber. The shop has a greatly expanded capacity for metalwork, chemical processing, and virtually every other aspect of aerospace artifact treatment. The museum's ability to conserve, preserve, and restore historic flying craft and related collections, which at Garber was already unmatched in the world, has reached an even higher standard at the Udvar-Hazy Center.

Among the first artifacts on the restoration floor will be the SB2C Helldiver, a World War II Navy dive-bomber. Don Engen flew Helldivers in the Pacific, and in tribute to him and at the request of his family, that aircraft will be restored and take its rightful place in the Udvar-Hazy Center.

In addition to the restoration facilities, the new building will hold the National Air and Space Museum's archives. The new facility will house more than 12,000 cubic feet of historic documents, 1.75 million photographs, and 14,000 films and video titles. It is one of the most significant aviation and space history archival collections to be found anywhere.

A state-of-the-art Collections Processing Unit will replace the one at Garber. This secure entry point for new museum artifacts will be staffed by specialists who will inspect, photograph, and add to the museum collections database; they will then store objects until they are needed for display. Also supporting the preservation of the museum artifacts is the Emil Buehler Conservation Laboratory, where innovative plans for preserving and restoring objects ensure that artifacts remain in excellent condition for future generations. The new | *to page 380* |

As an atmospheric flight-test vehicle, *Enterprise* has only dummy main engines. In this picture taken shortly after Space Hangar opening, the mock-ups of the two bulbous maneuvering rocket pods on either side of the tail have not yet been installed.

NO TRESPASSING
L. R. Willson & Sons, Inc.
STEEL ERECTORS

Designed to resemble an airport, the Hazy Center's exterior metal and blue tile finish changes mood with the rising and setting sun.

C O L L E C T I O N S

The Smithsonian's aeronautical model collection began early. In 1905, just two years after the Wrights' first flight, the Smithsonian collected three glider models designed by pioneer Octave Chanute.

M O D E L S

GEE-BEE R-1

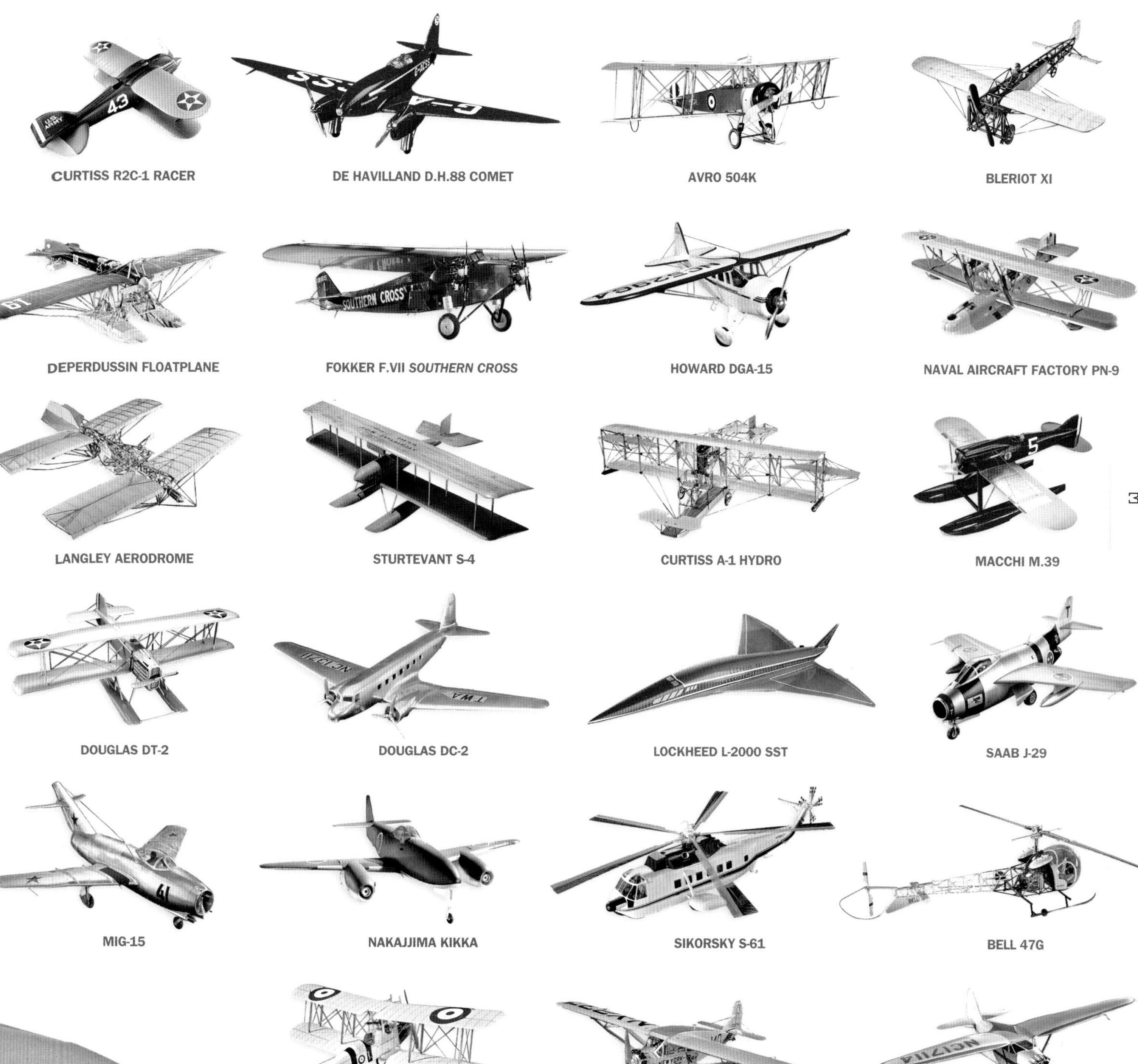

CURTISS R2C-1 RACER

DE HAVILLAND D.H.88 COMET

AVRO 504K

BLERIOT XI

DEPERDUSSIN FLOATPLANE

FOKKER F.VII *SOUTHERN CROSS*

HOWARD DGA-15

NAVAL AIRCRAFT FACTORY PN-9

LANGLEY AERODROME

STURTEVANT S-4

CURTISS A-1 HYDRO

MACCHI M.39

DOUGLAS DT-2

DOUGLAS DC-2

LOCKHEED L-2000 SST

SAAB J-29

MIG-15

NAKAJJIMA KIKKA

SIKORSKY S-61

BELL 47G

ROYAL AIRCRAFT FACTORY B.E.-2C

BELLANCA WB-2 *MISS COLUMBIA*

STINSON SR-10 RELIANT

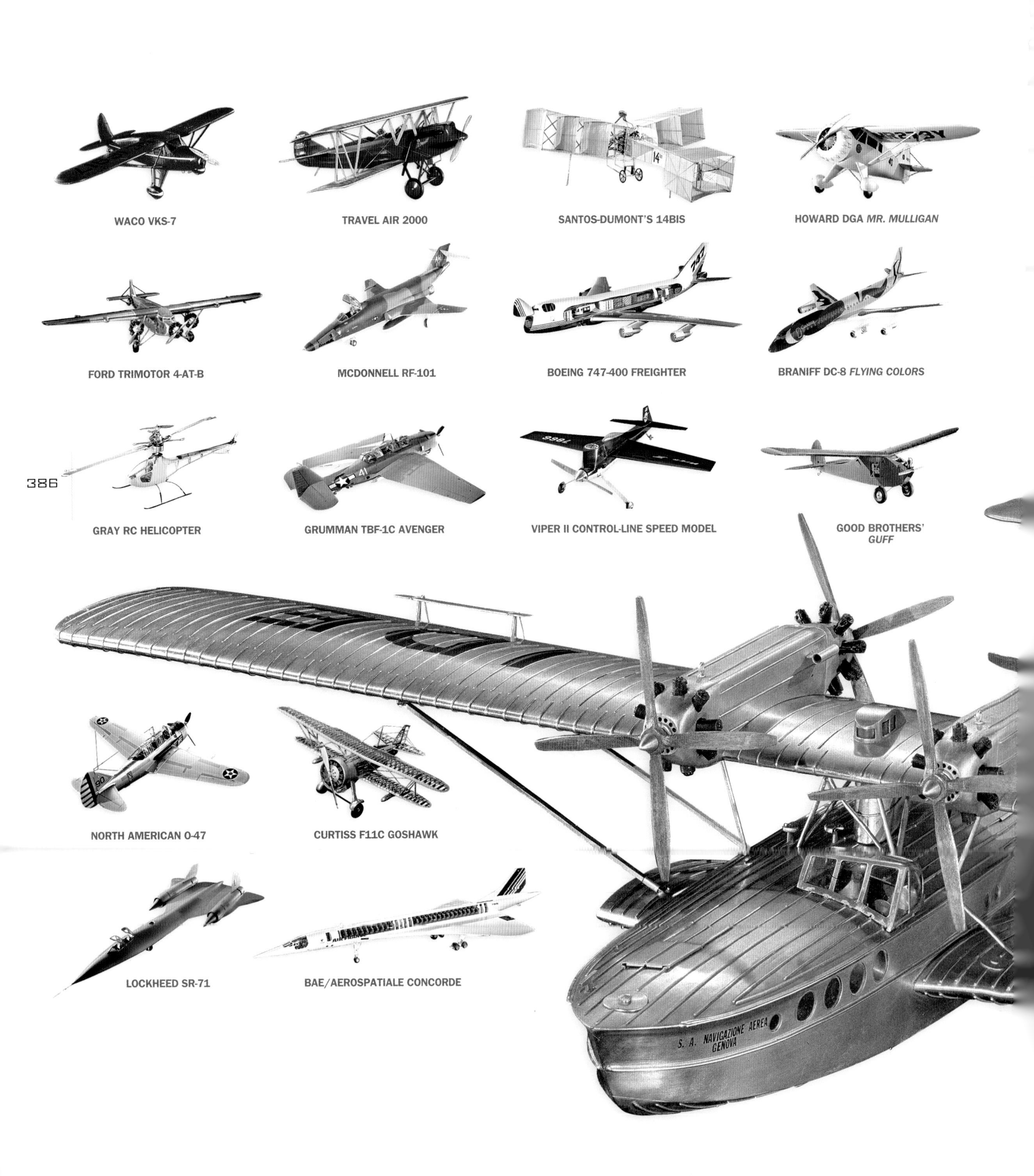

WACO VKS-7

TRAVEL AIR 2000

SANTOS-DUMONT'S 14BIS

HOWARD DGA *MR. MULLIGAN*

FORD TRIMOTOR 4-AT-B

MCDONNELL RF-101

BOEING 747-400 FREIGHTER

BRANIFF DC-8 *FLYING COLORS*

GRAY RC HELICOPTER

GRUMMAN TBF-1C AVENGER

VIPER II CONTROL-LINE SPEED MODEL

GOOD BROTHERS' *GUFF*

NORTH AMERICAN O-47

CURTISS F11C GOSHAWK

LOCKHEED SR-71

BAE/AEROSPATIALE CONCORDE

The man responsible for collecting most early aircraft, Paul Garber, was himself a modeler and avidly sought examples for display. The museum's collection of 3,000-plus artifacts includes those used to test new ideas, including the aerodromes Secretary Langley built as precursors to his Great Aerodrome. Wind tunnel models, from early carved mahogany examples to modern stainless-steel versions, have been collected as well. Aviation models have also served as promotional tools. Whether created to show off a manufacturer's latest design, to extol the virtues of air travel, or to provide ready-made desktop examples of favorite aircraft, display models make up the largest part of the collection. Perhaps the most popular ones, however, are those handbuilt by hobbyists. Garber organized contests, and the museum today continues to collect static and flying recreational models.

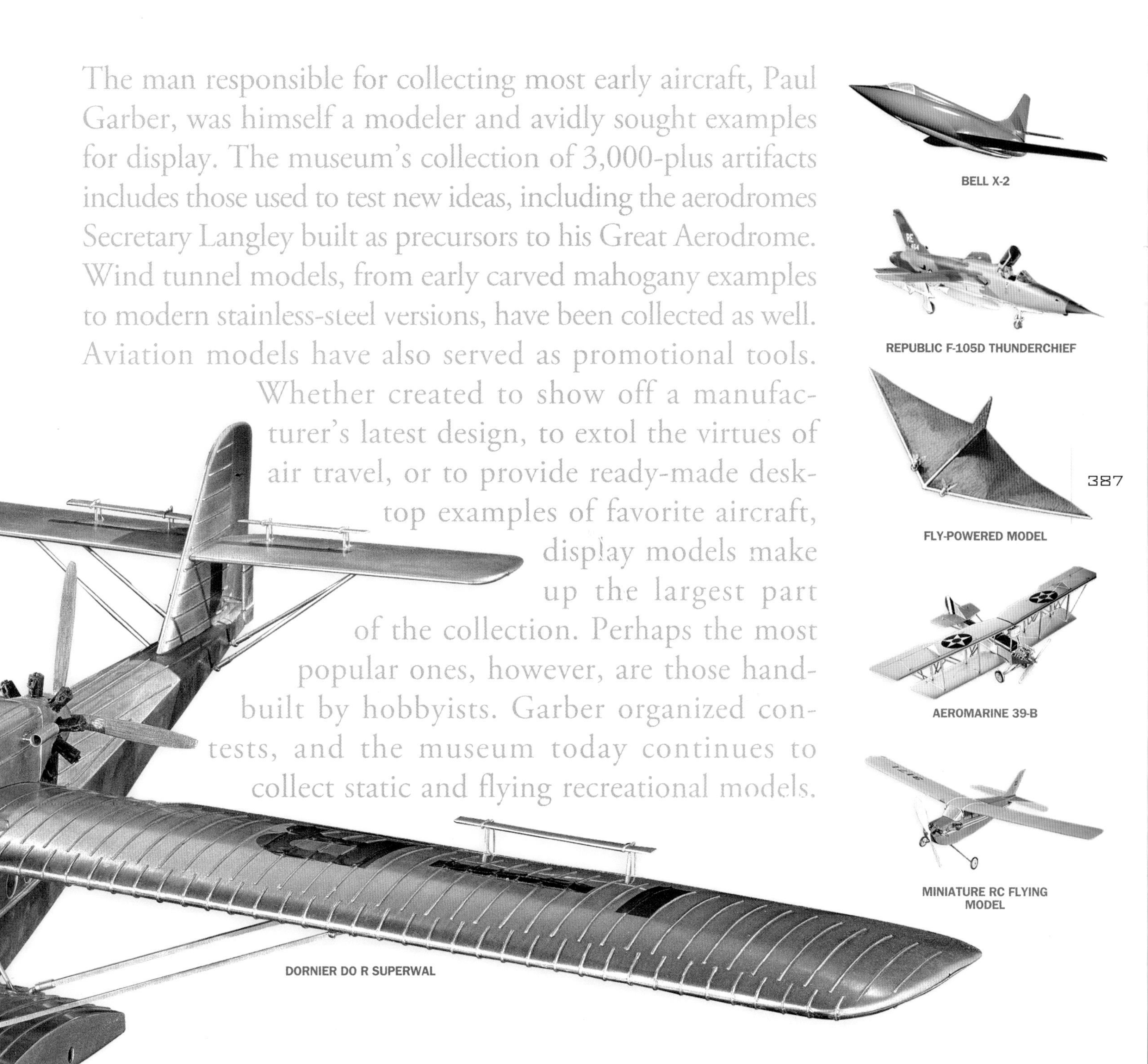

BELL X-2

REPUBLIC F-105D THUNDERCHIEF

FLY-POWERED MODEL

AEROMARINE 39-B

MINIATURE RC FLYING MODEL

DORNIER DO R SUPERWAL

John R. Dailey, director of the National Air and Space Museum, sits in the cockpit of a Marine Corps AV-8B Harrier II at Cherry Point, North Carolina, circa 1986. His call sign was "Zorro."

AFTERWORD

In the century from the Wright brothers' first flights at Kitty Hawk to the landing of remotely driven rovers on Mars, the visionaries, engineers, pilots, and businesspeople who have created the machines that soared over the Earth and reached into space have left a legacy of achievement that rivals any in human history. Paralleling that history, and at times intersecting and shaping it, has been the story of the National Air and Space Museum.

JOHN R. DAILEY

THE FUTURE OF THE NATIONAL AIR AND SPACE MUSEUM

As this book reveals, from Samuel Langley's early research into human flight to NASA accident investigations that borrowed space shuttle artifacts from our collection, the museum has played a direct role in the awe-inspiring enterprise of aerospace. From Paul Garber's single-minded dedication to building the museum's peerless collection, almost as history was happening, the museum has grown into the world's premier center for aviation and space artifacts, aerospace historical research, and the exhibition of aerospace science and technology. The history of human flight and the history of the museum have walked shoulder to shoulder for the last 100 years.

As we now move into the second aviation century, fascinating questions and challenges are before us. Will we return to the moon in the near future? Will we leave footprints farther out into the solar system on a human mission to Mars? Will our commercial air transport system again make everyday passenger travel economically viable and pleasant? How will the use of remotely piloted vehicles alter and enhance the application of military airpower? What greater understanding of the universe will future generations of space telescopes and robotic probes reveal? From what new exotic materials will we build our aircraft and spacecraft? What sorts of propulsion technology will power us through the atmosphere and across space? The answers to these and countless other questions will chart the history of aerospace for the generations of the 21st century.

Just as the development of the wondrous technology of flight will continue, so too will the growth of the National Air and Space Museum. From Paul Garber's first days at the Smithsonian, with every new storage room, every new exhibition hall, and every new building, the staff of the museum has always asked the question "What's next?" Every success at meeting the need to properly care for and display the national collection has been quickly followed by the challenge to find new and better ways to cope with an ever expanding collection.

The opening of Phase Two of the Udvar-Hazy Center, featuring the Mary Baker Engen Restoration Hangar and high-quality facilities to house the museum's archives and small-artifact collections, is the latest dramatic leap forward. The collections will have better care and the public will have greater access to them than ever. And, as ever, the museum will continue to look to the future. As the technology of aviation and spaceflight evolves in new and as yet unimagined directions, the museum will follow and chronicle those developments, and collect and preserve the artifacts and stories of this most exciting and influential of human endeavors.

The mission of the museum is to commemorate, educate, and inspire. No doubt many of the men and women who will shape aerospace in the 21st century will walk through its exhibitions, see the highlights of its collections, participate in its educational program, or read a book written by one of its scholars. It is our hope that the museum can provide the spark of interest and enthusiasm that will launch young people toward achieving great things in the field. As has been our tradition and our welcome responsibility, the Smithsonian National Air and Space Museum will continue to grow so that the artifacts created by these 21st-century aerospace pioneers will inspire other generations to come.

C O N T R I B U T O R S

JOHN H. GLENN, JR., made aviation history when, on February 20, 1962, he rode into space atop an Atlas rocket and piloted the *Friendship 7* spacecraft around the globe three times, becoming the first American to orbit Earth. As a Marine pilot, Glenn had flown 149 missions during World War II and the Korean War, earning the Distinguished Flying Cross six times. He was one of the first seven astronauts in the U.S. space program. In 1974, Glenn was elected as U.S. Senator from Ohio, a position he held until 1997. A year later, NASA invited him to rejoin the space program. On October 29, 1998, he became the oldest human ever to venture into space.

GEN. JOHN R. DAILEY (USMC, Ret.) is the director of the National Air and Space Museum. Prior to taking the helm, General Dailey was the associate deputy administrator of the National Aeronautics and Space Administration (NASA), where he led that agency's restructuring activities. He served 36 years in the U.S. Marine Corps, where he attained the rank of general and retired in 1992 while serving as assistant commandant. He is a pilot with more than 7,000 hours in numerous aircraft and helicopters. During two tours in Vietnam, he flew 450 combat missions. General Dailey has numerous decorations for his service in the Marine Corps and NASA. He is former national commander of the Marine Corps Aviation Association and a member of the Early and Pioneer Naval Aviators Association (Golden Eagles) and a number of other national and international organizations and boards.

MICHAEL J. NEUFELD is chair of the Space History Division of the National Air and Space Museum, Smithsonian Institution. Born and raised in Canada, he has four history degrees, including a Ph.D. from Johns Hopkins University in 1984. Dr. Neufeld has written three books, *The Skilled Metalworkers of Nuremberg* (1989); *The Rocket and the Reich* (1995), which won two book prizes; and *Von Braun* (2007), which has won three awards. He has also edited two others, *Planet Dora* (by Yves Béon) and *The Bombing of Auschwitz* (with Michael Berenbaum). He has appeared on numerous television and radio programs, notably on the History Channel, C-SPAN, PBS, and NPR, as well as the BBC, the German ZDF, and other foreign outlets.

ALEX M. SPENCER is curator of British aircraft and military flight matériel at the National Air and Space Museum. Born in Bellefonte, Pennsylvania, he attended the Pennsylvania State University at University Park, graduating with a B.A. in May 1987 and an M.A. in May 1990, both in history. In August 2008, he completed his Ph.D. in history at Auburn University. He joined the Aeronautics Division of NASM in April 1990.

MARK AVINO is chief of photographic services in the Office of Communications of the National Air and Space Museum. In addition to a wide variety of photography undertaken at the museum, he has also contributed to numerous publications and books on the museum's collections, including *In the Cockpit: Inside 50 History-Making Aircraft; At the Controls: The Smithsonian National Air and Space Museum Book of Cockpits; Spacesuits: The Smithsonian National Air and Space Museum Collection;* and *In the Cockpit II: Inside History-Making Aircraft of World War II.*

TOM D. CROUCH is senior curator of aeronautics at the National Air and Space Museum. A Smithsonian employee since 1974, he has held a variety of curatorial and administrative posts at both the museum and the National Museum of American History. He holds a B.A. from Ohio University, an M.A. from Miami University, and a Ph.D. from the Ohio State University. In 2002, the Wright State University awarded him a doctorate of humane letters. He is the author or editor of more than 15 books, most on aspects of the history of flight. His book *The Bishop's Boys: A Life of Wilbur and Orville Wright* won a 1989 Christopher Prize. He has also won book prizes offered by the American Institute of Aeronautics and Astronautics and the Aerospace Writers Association.

DIK A. DASO is curator of modern military aircraft at the National Air and Space Museum. He has curated exhibits at the Udvar-Hazy Center and NASM and is co-curator for The Price of Freedom: Americans at War, a permanent exhibition at the National Museum of American History. He has contributed chapters to several books and been the author of six. His biography *Hap Arnold and the Evolution of American Airpower* earned the 2001 American Institute of Aeronautics and Astronautics History Manuscript Award. A retired U.S. Air Force lieutenant colonel, he taught history at the USAF Academy and was chief of Air Force doctrine at Headquarters Air Force, Pentagon. He flew RF-4C Phantoms, F-15 Eagles, and T-38 Talon supersonic trainers, accumulating 2,750 flight hours during his military career. He holds a B.S. from the U.S. Air Force Academy and an M.S. and Ph.D. from the University of South Carolina.

MARILYN F. GRASKOWIAK is the chair of the National Air and Space Museum Archives. She received her undergraduate and graduate degrees in history from the University of Nebraska. After 20 years of teaching and managing the archives of a community of religious women, she decided to pursue a full-time career in archives collections management. Graskowiak joined the National Air and Space Museum Archives staff in 1992 and became chair of the division in 2005. The Archives Division will move to the Udvar-Hazy Center in 2011.

MELISSA A. N. KEISER is chief photo archivist for the National Air and Space Museum Archives. She joined the museum in 1985 as a copy photographer on the museum's archival videodisc project; 25 years later, she cares for more than two million images—one of the Smithsonian's largest collections of photography. She has used her image research talents in support of numerous museum publications and projects, including the exhibits America by Air (open 2007) and Pioneers of Flight (open 2010) and the books *The Smithsonian National Air and Space Museum Book of Flight* (2001) and *The Best of the National Air and Space Museum* (2006). She is co-author of *The Legacy of Flight: Images from the Archives of the Smithsonian National Air and Space Museum* (2010). Her subject specialties include early American aviation photographers, pre–World War I French aviation, and aircraft nose art.

ERIC F. LONG has been a senior photographer with the Smithsonian Institution since 1983. He began his photographic career at the Museum of American History and was then detailed to the National Air and Space Museum in 1996. His responsibilities include documenting collections, exhibitions, and events for museum research and publications. Long has participated in historical documentation of three presidential Inaugurations and an oral history of southern farmers in South Carolina. His work has been published in magazines such as *Air&Space/Smithsonian* and *Smithsonian* and a number of books, including *On Miniature Wings; Star Wars—The Magic of Myth; At the Controls; In the Cockpit: Inside 50 History-Making Aircraft; Sputnik: 50 Years of the Space Age;* and *In the Cockpit II: Inside History-Making Aircraft of World War II.*

TED A. MAXWELL is senior scientist at the Center for Earth and Planetary Studies of the Smithsonian National Air and Space Museum, where he is engaged in research to understand better the role of climate change in modifying land surfaces on Earth and Mars. He joined the staff of the museum in 1976 and has served as scientist, chair of the Center for Earth and Planetary Studies, senior adviser for science, and associate director for collections and research. He has been a principal investigator in several of NASA's research programs, is a past chair of the Planetary Geology Division of the Geological Society of America, and is a fellow of the American Association for the Advancement of Science and the Geological Society of America. His research concentrates on landscape evolution on Mars and on Earth, particularly Egypt and Sudan, where he has been conducting field studies since 1978.

DANE A. PENLAND is senior photographer for the National Air and Space Museum's Steven F. Udvar-Hazy Center. Throughout his many years at the Smithsonian, Penland has photographed many diverse topics, from full-size locomotives to microphotography of live larvae, and from the Supreme Court justices to the Hope Diamond in photographs that have become world famous. He has published numerous posters and has contributed to Institution publications, including *Reflections on the Wall: Vietnam Memorial; Suiting Everyone; Crossroads of Continents; American Clocks; The American Presidency; After Sputnik;* and *America's Hangar,* as well to his own book, *The National Gem Collection.*

DOMINICK A. PISANO is a curator in the Aeronautics Division at the National Air and Space Museum. During the more than 30 years Pisano has spent at the museum, he has been involved in numerous research and writing projects, as well as collections management, exhibitions, and service to the public. He received his Ph.D. in American studies from George Washington University in 1988. He lives in Alexandria, Virginia.

F. ROBERT VAN DER LINDEN is the chairman of the Aeronautics Division of the National Air and Space Museum. He is also curator of air transportation and special purpose aircraft. He is the responsible curator for the Milestones of Flight gallery, the Hall of Air Transportation, and the Golden Age of Flight gallery, and is the author of seven books, including *The Nation's Hangar: The Aircraft Collection of the Steven F. Udvar-Hazy Center; Airlines and Airmail: The Post Office and the Birth of the Commercial Aviation Industry;* and *The Boeing 247: The First Modern Airliner.* A native of Washington, D.C., van der Linden holds a B.A. in history from the University of Denver and an M.A. and Ph.D. in modern American, business, and military history from George Washington University, and he is a member of Phi Beta Kappa. He started his career at the Smithsonian as a volunteer in 1975.

SOURCES AND FURTHER READING

CHAPTER 1: FLIGHT AND THE SMITHSONIAN
BY TOM D. CROUCH

Thomas Coulson, *Joseph Henry: His Life and Work* (Princeton, N.J.: Princeton University Press, 1950) remains the best single-volume biography of the first Smithsonian Secretary. *The Papers of Joseph Henry,* vols. 1-8 (Washington, D.C.: Smithsonian Institution Press, 1972–1998) and vols. 9-12 (Washington, D.C.: Smithsonian Institution in association with Science History Publications, 2002–2007) are an invaluable resource and contain many of the documents quoted in this chapter. Tom D. Crouch, *Eagle Aloft: Two Centuries of the Balloon in America* (Washington, D.C.: Smithsonian Institution Press, 1983) offers fuller information on the work of John Wise, T. S. C. Lowe, Solomon Andrews, and others. E. F. Rivinius and E. M. Youssef, *Spencer Baird of the Smithsonian* (Washington, D.C.: Smithsonian Institution Press, 1992) and Ellis Yochelson, *Charles Doolittle Walcott, Paleontologist* (Kent, Ohio: Kent State University Press, 1998) are the best sources of information on those Smithsonian Secretaries.

For Samuel Langley and the Aerodromes, see Tom D. Crouch, *A Dream of Wings: Americans and the Airplane, 1875–1905* (New York: W.W. Norton, 1981); J. Gordon Vaeth, *Langley: Man of Science and Flight* (New York: Ronald Press, 1966); John D. Anderson, Jr., *A History of Aerodynamics: And Its Impact on Flying Machines* (Cambridge: Cambridge University Press, 1997); Samuel P. Langley, *Experiments in Aerodynamics* (Washington, D.C.: Smithsonian Institution, 1891); and Charles M. Manly, ed., *Langley Memoir on Mechanical Flight* (Washington, D.C.: Smithsonian Institution, 1911).

INTERCHAPTER 1.5: CAPABLE OF FLIGHT? THE WRIGHT-SMITHSONIAN CONTROVERSY,
BY TOM D. CROUCH

For a fuller account of the Wright-Smithsonian controversy, see: Tom D. Crouch, "Capable of Flight: The Saga of the 1903 Airplane," in Amy Henderson and Adrienne L. Kaeppler, eds., *Exhibiting Dilemmas: Issues of Representation at the Smithsonian* (Washington, D.C.: Smithsonian Institution Press, 1997), pp. 92-116.

CHAPTER 2: BUILDING A COLLECTION
BY F. ROBERT VAN DER LINDEN

Most of the material for this chapter came from the Smithsonian Institution Archives, which house the early records of the National Air and Space Museum and its predecessor organizations. Of particular use were Records Units 84, 162, 408, and 537. The most useful file contained a lengthy, detailed 1974 oral history of Paul Garber, which is located in RU 9592. The museum archives also contains a good hanging file that preserves numerous newspaper clippings for this period. Secondary sources consulted include F. Robert van der Linden, *The Nation's Hangar: The Aircraft Collection of the Steven F. Udvar-Hazy Center* (Washington, D.C.: Smithsonian Institution, National Air and Space Museum in association with Howell Press, Charlottesville, Va., 2004); and Dave Dooling, "History of the National Air and Space Museum," *Spaceflight* (July-August 1976), 249-262.

INTERCHAPTER 2.5: ROBERT GODDARD AND THE SMITHSONIAN BY TOM D. CROUCH

The two best biographies of Robert Goddard are: David A. Clary, *Rocket Man: Robert H. Goddard and the Birth of the Space Age* (New York: Hyperion, 2003) and Milton Lehman, *This High Man: The Life of Robert H. Goddard* (New York: Farrar, Straus, 1963), republished as *Robert H. Goddard: Pioneer of Space Research* (New York: Da Capo Press, 1988). Relevant correspondence can be found in Esther C. Goddard and G. Edward Pendray, eds., *The Papers of Robert H. Goddard,* 3 vols. (New York: McGraw-Hill, 1970).

CHAPTER 3: THE LONG ROAD TO A NEW MUSEUM
BY DOMINICK A. PISANO

Tom D. Crouch, *Rocketeers and Gentlemen Engineers: A History of the American Institute of Aeronautics and Astronautics . . . And What Came Before* (Reston, Va.: American Institute of Aeronautics and Astronautics, 2006), pp. 174-176; Dave Dooling, "History of the National Air and Space Museum" (see above, Chapter 2); Heather Ewing and Amy Ballard, *A Guide to Smithsonian Architecture* (Washington, D.C.: Smithsonian Books, 2008), pp. 104-109; S. Paul Johnston, "An Address Given Before the Washington Aero Club, Tuesday, 22nd April 1969, by S. Paul Johnston, Director, National Air and Space Museum, Smithsonian Institution (Concerning the Status of the National Air & Space Museum)," *AAHS Journal* (Summer 1969); Joanne M. Gernstein London, "A Modest Show of Arms: Exhibiting the Armed Forces and the Smithsonian Institution, 1945–1976" (Ph.D. dissertation, George Washington University, 2000), pp. 92-104; James Gilbert, "The Dream and the Junkyard," *Flying* (March 1969), 42-45, 79-80; Dominick A. Pisano, F. Robert van der Linden, and Frank H. Winter, *Chuck Yeager and the Bell X-1: Breaking the Sound Barrier* (Washington, D.C.: Smithsonian National Air and Space Museum in association with Harry N. Abrams, 2006), pp. 127-132; Hereward Lester Cook with James D. Dean, *Eyewitness to Space: Paintings and Drawings Related to the Apollo Mission to the Moon* (New York: Harry N. Abrams, 1971); Zachary M. Schrag, *The Great Society Subway: A*

History of the Washington Metro (Baltimore: Johns Hopkins University Press, 2006), pp. 22-24; F. Robert van der Linden, *The Nation's Hangar* (see above, Chapter 2).

INTERCHAPTER 3.5: THE NASA-NASM PARTNERSHIP BY MICHAEL J. NEUFELD

The NASA History Division at NASA Headquarters in Washington, D.C., has a file on the origins of the agreement, no. 14131. A little material is also found in SI Archives, Records Units 162 (for NACA), 306, and 348. On the Saturn Vs see: for the Kennedy Space Center, Frank H. Winter and Scott Wirz, "Saturn V Reborn—A Giant Restoration," *Journal of the British Interplanetary Society* 50 (1997), 169-172, popularized as "Saturn V Rising," *Air & Space Smithsonian* 11, no. 5 (December 1996/January 1997), 28-35, and for the U.S. Space and Rocket Center, Frederick I. Ordway III, Brenda M. Carr, and Irene Willhite, "Restoring a Giant," *Launch Magazine* (January–February 2008), 38-45, 79. On the Johnson Space Center Saturn V restoration there is nothing except newspaper articles and websites: see, e.g., *www.collectspace.com/*. NASM Space History Division has restoration records. Curator Valerie Neal provided a valuable chronology of the shuttle *Enterprise.*

CHAPTER 4: THE WORLD'S MOST POPULAR MUSEUM BY TED A. MAXWELL AND TOM D. CROUCH

A snapshot of the National Air and Space Museum soon after its opening is documented in C. D. B. Bryan, *The National Air and Space Museum* (New York: Harry N. Abrams, 1979). Rich with anecdotes as well as information on the aircraft stored there in the 1980s, Walter Boyne's *The Aircraft Treasures of Silver Hill* (New York: Rawson Associates, 1982) is an excellent resource. An updated history of the Paul E. Garber Facility is presented by F. Robert Van der Linden in *The Nation's Hangar* (see above, Chapter 2).

Criticisms of museum exhibits during the first two decades of the mall museum were written by Michal McMahon, "The Romance of Technological Progress: A Critical Review of the National Air and Space Museum," *Technology and Culture* 22 (1981), 281-296, and Samuel A. Batzli, "From Heroes to Hiroshima: The National Air and Space Museum Adjusts Its Point of View," *Technology and Culture* 31 (1990), 830-837. Director Martin Harwit offered his version of the *Enola Gay* exhibit episode in *An Exhibit Denied: Lobbying the History of* Enola Gay (New York: Copernicus, 1996), while Tom Crouch provided another view in "Aerospace Museums—A Question of Balance," *Curator: The Museum Journal* 50 (2007), 19-32. Criticisms of museum collections care are documented by the General Accounting Office (GAO) in *Smithsonian Institution—Better Care Needed for National Air and Space Museum Aircraft* (Washington, U.S. General Accounting Office, GAO/GGD-96-9, 1995). For the story of the Space Race gallery and its rare collection of artifacts, see Martin J. Collins, *Space Race: The U.S.-U.S.S.R. Competition to Reach the Moon* (San Francisco: Pomegranate, 1999). Donald S. Lopez's memoirs are *Into the Teeth of the Tiger* (Washington, D.C.: Smithsonian Institution Press, 1997) and *Fighter Pilot's Heaven: Flight Testing the Early Jets* (Washington, D.C.: Smithsonian Institution Press, 1995).

INTERCHAPTER 4.5: THE SMITHSONIAN AND THE *ENOLA GAY* BY DIK A. DASO AND TOM D. CROUCH

On the B-29 and the airplane, see Norman Polmar, *The* Enola Gay: *The B-29 That Dropped the Atomic Bomb on Hiroshima* (Washington, D.C.: Brassey's, 2004); Richard H. Campbell, *They Were Called Silverplate* (Tucson, Ariz.: Becam Press, 2003); Jacob Vander Meulen, *Building the B-29* (Washington, D.C.: Smithsonian Institution Press, 1995); Dik A. Daso, *Hap Arnold and the Evolution of American Airpower* (Washington, D.C.: Smithsonian Institution Press, 2000). For the 1994-95 exhibit controversy, see: Edward T. Linenthal and Tom Engelhardt, eds., *History Wars: The* Enola Gay *and Other Battles for the American Past* (New York: Metropolitan, 1996); Philip Nobile, ed., *Judgment at the Smithsonian* (New York: Marlowe, 1995) (the original Crossroads script); Michael J. Hogan, ed., *Hiroshima in History and Memory* (Cambridge: Cambridge University Press, 1996); Martin Harwit, *An Exhibit Denied,* and Tom D. Crouch, "Aerospace Museums—A Question of Balance" (see above, Chapter 4); for the Air Force Association point of view, see *www.afa.org/media/enolagay/.*

CHAPTER 5: THE NATIONAL AIR AND SPACE MUSEUM SPREADS ITS WINGS BY DIK A. DASO

Source material for this chapter comes mainly from official documents and correspondence between the major players involved with the development of the center. Many thanks to Lin Ezell for her work as the project manager for the construction of the Udvar-Hazy Center and also her authorship of the book *Building America's Hangar: The Design and Construction of the Steven F. Udvar-Hazy Center* (Washington, D.C.: Smithsonian National Air and Space Museum, 2004). Lin kept excellent files and utilized them in great detail in recounting the many events surrounding the construction and opening of the Udvar-Hazy Center in December 2003. Much of the background information for this chapter finds its roots in her book, Chapters 1-4. Oral history interviews with Leo Schefer and Tom Moore conducted by Tom Pumpelly at the Udvar-Hazy Center in September 2009 provided the corporate memory from the "Gang of Four" and the ASH Council, complete with an extensive documentary record of the period. Gen. Henry H. "Hap" Arnold's National Air Museum–related papers are kept in the museum Archives (Ramsey Room). For the most recent events, both museum and center staff contributed their recollections and anecdotes. Particularly helpful were Linda Hicks, Special Events; Dick Gentz, Phase One and Two Special Assistant to the Director; and Margie Natalie and Doug Baldwin, UHC Education. The UHC guidebook, Dik A. Daso, ed., *America's Hangar,* 3rd ed. (Washington, D.C.: NASM, 2010), has been expanded and revised twice since 2003.

I L L U S T R A T I O N C R E D I T S

Abbreviations: B = bottom; **C** = center; **L** = left; **R** = right; **T** = top. **NASM** = National Air and Space Museum; **SI** = Smithsonian Institution; **NASA** = National Aeronautics and Space Administration; **MA** = Mark Avino; **EF** = Eric F. Long; **DH** = Dale E. Hrabak; **DP** = Dane A. Penland. *Note:* all images identified by artifact number only (numbers prefixed "A") are courtesy of NASM Collections Management.

Front cover: DP, SI 2005-6521; **1, 2-3**: EF, NASM 9A07638; **5**: SI 92-6815; **6L**: SI 88-6821; **6R**: SI 2001-6507-12; **9**: EF, SI 2002-19481; **10-01**: SI 2003-6592; **12-13**: SI 91-14704; **14-15**: EF, SI 2006-7805; **16-17**: EF, WEB10104-2004. **Chapter 1. 18-19:** SI 2003-31614; **22TL:** EF, SI 2005-15502; **22TR:** DH, SI 79-13575; **22CL:** EF, SI 2002-19481; **22BL:** DH, SI 2002-14833; **23T:** DP, SI 80-2081; **23B:** DH, SI 79-4639; **24T:** MA, SI 86-12094; **24CL:** EF and MA, SI 98-15919; **24B:** MA, 91-7835; **24-25:** Richard B. Farrar, SI 79-4629; **25T:** DH, SI 81-14836; **25CR:** DP, NASM 9A08038; **25B:** DP, SI 2005-24666; **26:** SI 99-40777; **27:** SI 80-7745; **28:** National Portrait Gallery, Smithsonian Institution, NPG.64.10; **29:** MA, SI 2001-5358; **30-31:** SI 99-40780; **32:** NASM 7A47252; **33:** Library of Congress, LC-B8171-2348; **34:** Library of Congress, LC-DIG-cwpb-01563; **35:** A19290008000; **36-37:** SI 2002-16633; **38:** A18890001000; **39T:** SI 95-8647; **39C:** SI 95-8649; **39B:** SI 95-8652; **43:** SI 91-1455; **44:** NASM A-10983-B; **45:** NASM A-10983-E; **46:** DH, SI 98-15303; **49:** SI 2002-15561; **50-51:** NASM A-53902; **52T:** SI 2003-31623; **52C:** NASM A-5594; **52B:** NASM A-23761; **53:** NASM A-12525; **54:** A19820736000; **55:** Garnet Jex, *Langley Model #5 and Houseboat* (1933), NASM Art Collection, SI 96-15732; **56-57:** NASM A-12566; **58:** A19190004000; **59:** SI 2002-16636; **60:** NASM A-33416-A; **61:** SI 2002-16637; **62:** EF, SI 2007-1298 (A19762019000); **63:** Smithsonian Institution Archives, NASM 9A02224; **64-65:** SI 2003-30819; **68:** Library of Congress, LC-W85-114; **69:** EF, SI 97-16653 (A19640054000); **70:** A19090006000; **71:** Library of Congress, LC-W85-117; **73:** EF, SI 2005-20387; **75:** NASM A-1006-B; **76:** A19520060000; **77T:** SI 2002-10626; **77B:** SI 2009-12043; **78-79:** SI 85-10844; **80T:** SI 99-40427; **80B:** SI 96-16169; **81:** SI 96-16173; **83:** NASM 9A07347; **84-85:** NASM 9A06058; **86:** A19500182002; **87:** NASM 9A05084; **88-89:** SI 91-3480. **Aero Propulsion**, *pages 90-93, all photos by NASM.* **90**: DP, NASM 9A08053; **91** (T-B, L-R) **row 1** (EF): SI 2002-20999, SI 2008-14043, A19500082000, SI 2008-13911; **91 row 2** (EF): SI 2003-28543-2, SI 2003-28563, SI 2003-28955, SI 2003-28541-8; **91 row 3**:A19730230000, (EF) SI 2008-13913, A19660150000, SI 2009-30499; **91 row 4**: (EF) SI 2003-28958, A19900069000; **92 row 1**: (EF) SI 2003-28970, A19520106000, A19620052000, (EF) SI 2003-28960; **92 row 2**: A19420027000, A19640017000, A19610135000, (EF) SI 2008-13921; **92 row 3**: (DP) SI 2006-20989, (EF) NASM 9A08058, A19880241000, A19490009000; **92-93**: EF, SI 2003-28542; **93T**: A19850445000; **93C**: DP, SI 2006-4029; **93B**: A19520103000. **Interchapter 1.5. 94**: SI 89-20604; **96**: NASM A-49608-E; **97**: SI 2002-15560; **98**: SI 2004-1998; **99T**: NASM 9A01255; **99B**: SI 92-16999; **101**: SI 2005-30950. **Chapter 2. 102-103:** SI 2009-31186; **106TL:** DP, NASM 9A08059; **106BL:** EF, SI 2001-1890; **106-107:** SI 98-16295; **107TL:** DP, NASM 9A03171; **107TR:** SI 2005-29882; **107CR:** EF, SI 2009-7961; **107BR:** EF, SI 2005-22902; **108TL:** EF, SI 97-16097; **108CL:** DP, SI 2006-853; **108BL:** Richard B. Farrar, SI 74-4295; **108R:** EF, SI 97-16096; **109TL:** DP, SI 2005-6566; **109TR:** SI 2005-35239; **109C:** DP, SI 2004-18376; **109BL:** DP, SI 2006-28363; **109BR:** DH, SI 80-17160; **110-111:** SI 2002-10653; **112:** A19610155000; **113T:** NASM 00196568; **113C:** stilll from "Conquest of the Air" (1936), London Film Productions, SI 90-9554; **113B:** NASM A-43735-A; **114-115:** SI 75-5034; **116T:** courtesy American Airlines, SI 93-1066; **116B:** courtesy American Airlines, NASM 00138566; **117:** SI 75-7024; **118:** Alex M Spencer, A19970523000; **119:** SI 81-876; **120:** MA, SI 97-15335; **122T:** SI 2001-13012; **122B:** SI 78-4653; **123:** NASM 9A06448; **124:** SI 99-40459 (A19880404000); **125:** NASM A-45236-A; **126BL:** NASM 9A06832; **126BR:** NASM 9A06845; **126-127:** NASM 00040439; **127BL:** NASM 9A06858; **127BR:** NASM 9A06862; **128-129:** NASM 9A05605; **132-133:** Geoffrey Watson, *NC-4*, NASM Art Collection (A19720256000), SI 90-335; **134-135:** NASM 7A12322; **136-137:** NASM 9A06086; **138:** A19540005000; **139:** NASM A-42474; **140T:** NASM USAF-3841AC; **140C:** NASM A-746; **140B:** SI 2002-12192; **141:** SI 95-2379 (A19670152000); **142:** A19360036000; **143T:** NASM 9A00249; **143B:** SI 91-3702; **144-145:** photo by *The Washington Evening Star* , May 1, 1936 edition, reproduced courtesy The Washington Post Co. (SI 89-11607); **146-147:** SI 80-444; **148-149:** NASM 9A05082; **150:** SI 2000-9524; **151T:** NASM A-59835-I; **151B:** SI 97-17486; **152:** EF, SI 97-16073; **154-155:** NASM 9A06449; **156:** A19730887000; **157:** SI 97-17028; **158-159:** Edward J. Steichen, NASM 9A03960; **160-161:** SI 85-7292; **161T:** NASM 7A33871; **161CT:** NASM A-4890-F; **161CB:** NASM A-45879-F; **161B:** SI 97-17475; **162:** Alfred T. Palmer/Library of Congress Prints and Photographs Division, LC-USW36-489; **163:** A19780399000; **165:** SI 2002-12355; **166:** NASM A-42868-B; **167:** A19470030000; **168-169:** SI 91-3701. **Posters**, *pages 170-173, all photos by NASM.* **170:** SI 98-20676; **171 (T-B, L-R) row 1:** A19980100000, SI 98-20032, SI 98-20034, SI 98-20047, SI 98-20049, SI 98-20664; **171 row 2:** SI 98-20616, SI 2003-4326, SI 98-20066, SI 98-20067, SI 98-20069, SI 98-20070; **171 row 3:** SI 98-20046, SI 98-20679, SI 98-20680, SI 98-20665, SI 98-20703, SI 98-20045; **171 row 4:** SI 99-15703, SI 98-20735, SI 98-20667, SI 98-20666, SI 98-20439, SI 98-20438; **172 row 1:** SI 98-20728, SI 98-20727, SI 98-20634, SI 98-20498, SI 2000-5226, SI 98-20430; **172 row 2:** SI 98-20429, SI 98-20416, SI 98-20412, SI 98-20409, SI 98-20404, SI 98-20101; **172 row 3:** SI 98-20695, SI 98-20696, SI 98-20697, SI 98-20698, SI 98-20700, SI 98-20701; **172 row 4:** SI 98-20073, SI 98-20079, SI 98-20092, SI 98-20083, SI 81-6659, SI 98-20528; **173:** SI 2003-5511-9. **Interchapter 2.5.**

174-175: I 84-8949; **176**: SI 95-8690; **177**: SI 94-12969; **178-179**: B. Anthony Stewart/National Geographic Image Collection, NGS Image ID: 971396. **Chapter 3 180-181:** NASM A-52236; **184T**: EL, SI 2000-9403; **184C**: EL, SI 2001-11467; **184BL**: DP, NASM 9A08061; **184BR**: DP, SI 2004-18354; **184-185**: DP, SI 2005-6275; **185TL**: EL, SI 2008-5559-2; **185TR**: DP, SI 2005-6540; **185B**: DP, NASM 9A08060; **186TL**: EL, SI 2005-24509; **186TR**: EL, SI 2006-25353; **186C**: EL, SI 2000-9346; **186BL**: EL, SI 2000-9371; **187TL**: EL, SI 98-16042; **187TC**: EL, SI 2005-22904; **187TR**: EL, SI 2002-591-07; **187BL**: EL, SI 99-15232; **187BR**: EL, SI 2005-20394; **188:** A19490049000; **189T:** NASM 9A00849; **189B:** SI 80-4550; **190-191:** SI 2006-28133; **192-193:** SI 2009-31392; **194:** A19540106000; **195T:** NASM 00194461; **195B:** NASM 00037828; **196T:** NASM A-42188; **196B:** NASM A-42188-A; **197:** EF, SI 2005-36194-05 (A19680478000); **198-199:** NASM 9A07586; **202:** A20090132000; **203T:** SI 86-14171; **203B:** SI 2003-25360; **204:** A19940068000; **205T:** NASM 00101898; **205B:** SI 2004-50580; **206:** SI 80-19927; **208:** SI 98-15403; **209:** EF and MA, SI 2001-134; **210:** NASM 9A08020; **211:** SI 2009-31455; **212-213:** SI 2009-31452; **214:** A20090052000; **215T:** copied from McKim, Mead and White drawing, NASM A-43307-A; **215CT:** NASM A-43339-G; **215CB:** NASM 00193861; **215B:** SI 72-10112; **216-217:** NASM 9A07495; **218-219:** NASM 00163626; **222:** DP, SI 2005-35469 (A20020366000); **223T:** NASM 7B21572; **223B:** NASA S62-00303; **224:** A20060573000; **225:** NASM 00159960; **226:** A19790559000; **227:** SI 73-4622; **228:** SI 2006-28132; **230:** NASA MR3-19; **231:** NASA MR3-45; **232:** A19850133000; **233:** SI 72-10569; **234-235:** SI 72-11326-2; **236:** SI 74-4117; **237:** A19791775000; **238-239:** SI 2009-31393; **240-241:** NASM 9A07396; **242-243:** NASM 9A07399; **244:** SI 2009-31389; **245:** SI 2000-1566; **246-247**: SI 2009-31387. **Insignia**, *pages 248-251, all images by Alex M Spencer.* **Interchapter 3.5. 252**: NASM 9A03111; **253:** NASA GPN-2006-000038; **255**: NASM 7B21889; **256**: Thomarios (photo no. S1996-5712-14), via NASM Space History (NASM 9A08029); **257**: Thomarios (photo no. S1996-9249-11), via NASM Space History (NASM 9A08022); **258**: DP; **259**: DP. **Chapter 4. 260-261:** EF, SI 2005-5013-6; **264TL**: EL, SI 2006-25314; **264C:** MA, NASM 9A02849; **264B**: EL, SI 2005-22898; **265L:** courtesy Breitling SA, SI 2006-21451; **265R**: EL, SI 2006-3297; **266T:** DP, SI 2005-24462; **266B:** DP, SI 80-3070; **267TR:** EL, SI 2008-4230; **267CL**: DP, NASM 9A08062; **267CR:** EL, SI 2008-4224; **267B**: EL, SI 2008-4223; **268-269:** SI 2009-31384; **270:** A20070066000; **271:** NASM 9A07636; **272-273:** NASM 9A07635; **274:** A19850088000; **275T:** SI 84-4714-8; **275B:** NASM 9A08056; **276-277:** Still from the IMAX film "To Fly!" (1976), produced by MacGillivray Freeman Films for Francis Thompson Inc. and the National Air and Space Museum, Smithsonian Institution, sponsored by Conoco, a subsidiary of DuPont (NASM 9A08039); **280-281:** courtesy NASA, SI 84-6698; **282:** EF, SI 2006-1138-07 (A19972740000); **283:** NASM 7B21094; **284:** NASM 9A06879; **285:** EF, SI 98-16287-7; **286-287:** MA, SI 2008-2239; **288:** EF, SI 2002-2216; **289T:** NASM 9A08057; **289B:** EF, SI 2001-1394; **290-291:** EF, SI 2000-9368; **292:** EF, SI 98-15573-5 (A19980134000); **293:** Howard Barbie, SI 85-3949; **294:** MA, SI 2006-11212; **296-297:** Doug Shane / Visions, NASM 9A02842; **298:** A19960513000; **299T:** MA, SI 85-17474; **299B:** NASM 9A04016; **300T:** SI 99-15433; **300B:** Imperial War Museum, London (photo OWIL. 64336), SI 75-15871; **301:** A20000450000; **302:** NASA GPN-2000-001064; **304:** NASA GPN-2000-000938; **305:** EF, SI 2001-6718; **306:** A19730875000; **307T:** SI 95-2480; **307B:** SI 85-9421-10; **310:** SI 97-16261-11 (A19950118000); **311T:** Jim Zimbelman, NASM Center for Earth and Planetary Studies (CEPS), NASM 9A08042; **311B:** EF, NASM 9A04021; **312-313:** courtesy of Dr. Ross Irwin, NASM Center for Earth and Planetary Studies (CEPS). **Space Suits,** *pages 314-316, all photos by MA.* **314**: SI 2004-10884; **315** (L-R, T-B) **row 1**:SI 2004-19149; SI 2005-26165, SI 2003-37180, SI 2003-37182, SI 2004-59968; **315 row 2**: SI 97-15263, SI 97-15265, SI 2002-32829, SI 2002-32836, SI 2003-27279; **315 row 3**: SI 2003-27305, SI 2003-27311, SI 2003-27315, SI 2003-27343, SI 2003-27345; **316 row 1**: SI 2003-27353, SI 2003-27355, SI 2003-27359, SI 2003-27375, SI 2003-37196; **316 row 2**: SI 2004-10867, SI 2004-10869, SI 2004-10874, SI 2004-10822, SI 2004-60883; **316 row 3**: SI 2004-25833, SI 2004-25835, SI 2004-55438, SI 2004-59974, SI 2005-14756; **317**: NASA S88-52685. **Interchapter 4.5. 318-319**: SI 99-42697; **320-321**: NASM 9A05310; **322**: NASM USAF-58188AC; **323**: EF and MA, SI 98-15873; **325**: DP. **Chapter 5. 326-327:** DP; **330TL**: MA, SI 2003-44812; **330B**: DP, SI 2006-18727; **331TL**: DP, NASM 9A08066; **331TR**: DP, NASM 9A08063; **331CR**: DP, NASM 9A08067; **331BR**: DP, NASM 9A08068; **332T**: DP, NASM 9A08064; **332C**: DP, SI 2006-24265; **332B**: DP, NASM 9A08065; **333TL**: A19860098000; **333TR**: DP, SI 2006-104; **333B**: DP, SI 2005-1520; **334-335:** Jeff Tinsley, NASM 9A08019; **336-337:** MA, SI 92-14116; **339:** NASM 9A01552; **341:** SI 99-15320; **342-343:** EF, NASM 9A00111; **344-345:** NASM 9A08045; **346T:** SI 2002-2573-34; **346B:** MA, NASM 9A01564; **347:** A19730264000; **348-349:** NASM 9A01565; **350:** DP, SI 2004-19462; **352:** SI 74-8669; **353T:** SI 2004-11294; **353B:** National Archives and Records Administration (80-G-345181), SI 2009-12037; **354:** A20040290038; **355:** SI 2004-58219; **356-357:** EF, SI 2004-18275-34; **360-361:** DP; **362:** A20020370000; **363T:** SI 2005-14079; **363B:** Carolyn Russo; **364T:** DP; **364B:** SI 2006-7537; **365:** A20060587000; **366-367:** DP, NASM 9A07639; **368:** A20010071000; **369:** DP, SI 2006-12604; **372:** Joint Strike Fighter Office, DoD, via Dik Daso; **372:** DP, NASM 9A03884; **373:** Dik Daso, NASM 9A08054; **374:** SI 2006-789-07 (A20050477000); **375:** SI 2004-40698; **376-377:** DP, SI 2004-58551; **380-381:** DP; **382-383:** DP. **Aircraft Models**, *pages 384-387, all photos by EF.* **384**: SI 94-2270; **385** (T-B, L-R), **row 1**: SI 94-2181, SI 94-2182, SI 94-2183, SI 94-2184; **385 row 2**: SI 94-2193, SI 94-2195, SI 94-2196, SI 94-2197; **385 row 3**: SI 94-2198, SI 94-2208, SI 94-2205, SI 94-2209; **385 row 4**: SI 94-2212, SI 94-2216, SI 94-2219, SI 94-2220; **385 row 5**: SI 94-2221, SI 94-2223, SI 94-2225, SI 94-2226; **385 row 6**: SI 94-2248, SI 94-2250, SI 94-2252; **386, row 1**: SI 94-2253, SI 94-2278, SI 94-2255, SI 94-2279; **386, row 2**: SI 94-2336, SI 94-2312, SI 94-2329, SI 94-2316; **386, row 3**: SI 94-2326, SI 94-2321, SI 94-2303, SI 94-2327; **386, row 4**: SI 94-2315, SI 94-2296; **386, row 5**: SI 94-2311, SI 94-2310; **386-387**: SI 94-2335; **387**, T-B: SI 94-2287, SI 94-2299, SI 94-2288, SI 94-2298, SI 94-2301. **Afterword**: **388**: NASM 9A07868. **Back cover**: MA, SI 2004-10884.

Boldface indicates illustrations.
If illustrations are included within a page span, the entire span is **boldface.**
NASM stands for National Air and Space Museum. SI stands for Smithsonian Institution.

L

M

N

O

P

R

PUBLISHED BY THE NATIONAL GEOGRAPHIC SOCIETY

John M. Fahey, Jr., President and Chief Executive Officer
Gilbert M. Grosvenor, Chairman of the Board
Tim T. Kelly, President, Global Media Group
John Q. Griffin, Executive Vice President; President, Publishing
Nina D. Hoffman, Executive Vice President;
President, Book Publishing Group

PREPARED BY THE BOOK DIVISION

Barbara Brownell Grogan, Vice President and Editor in Chief
Marianne R. Koszorus, Director of Design
Carl Mehler, Director of Maps
R. Gary Colbert, Production Director
Jennifer A. Thornton, Managing Editor
Meredith C. Wilcox, Administrative Director, Illustrations

STAFF FOR THIS BOOK

Susan Tyler Hitchcock, Project Editor
Bronwen Latimer, Illustrations Editor
Sam Serebin, Art Director
Judith Klein, Production Editor
Lewis Bassford, Production Project Manager
Marshall Kiker, Illustrations Specialist
Al Morrow, Design Assistant

MANUFACTURING AND QUALITY MANAGEMENT

Christopher A. Liedel, Chief Financial Officer
Phillip L. Schlosser, Vice President
Chris Brown, Technical Director
Nicole Elliott, Manager
Rachel Faulise, Manager

The National Geographic Society is one of the world's largest nonprofit scientific and educational organizations. Founded in 1888 to "increase and diffuse geographic knowledge," the Society works to inspire people to care about the planet. It reaches more than 325 million people worldwide each month through its official journal, *National Geographic,* and other magazines; National Geographic Channel; television documentaries; music; radio; films; books; DVDs; maps; exhibitions; school publishing programs; interactive media; and merchandise. National Geographic has funded more than 9,000 scientific research, conservation and exploration projects and supports an education program combating geographic illiteracy.
For more information, visit nationalgeographic.com.

For more information, please call 1-800-NGS LINE (647-5463) or write to the following address:

National Geographic Society
1145 17th Street N.W.
Washington, D.C. 20036-4688 U.S.A.

Visit us online at www.nationalgeographic.com

For information about special discounts for bulk purchases, please contact National Geographic Books Special Sales: ngspecsales@ngs.org

For rights or permissions inquiries, please contact National Geographic Books Subsidiary Rights: ngbookrights@ngs.org

ISBN: 978-1-4262-0653-5

Printed in China

10/RRDS/1

LIVINGSTON PUBLIC LIBRARY